W9-DEP-069

AFRICA, THE DEVASTATED CONTINENT?

MONOGRAPHIAE BIOLOGICAE

Editor

J. ILLIES

Schlitz

VOLUME 26

DR. W. JUNK b.v. PUBLISHERS THE HAGUE 1975

AFRICA, THE DEVASTATED CONTINENT?

Man's impact on the ecology of Africa

by

ANTOON DE VOS

DR. W. JUNK b.v. PUBLISHERS THE HAGUE 1975

To VANESSA without whose help this book would never have come to fruition.

ISBN 90 6193 078 2

Cover Design M. Velthuijs, The Hague
Printed in the Netherlands

CONTENTS

	Foreword	11
	Introduction	13
I	The African environment	15
II	Ecological zones of Africa	26
III	Man as an environmental agent	102
IV	Specific environmental problems	128
V	Problems, needs and potentials in land use	163
VI	Planning for the future	215
	Literature cited	227
	Index	231

CHAPTER'S CONTENTS

I. **The African environment** 15
Geology 15
Geography. 15
Phytogeography 16
Zoogeography 16
Climate 19
Soils 20
Vegetation types 22
Woodlands, savannas and steppes 22
Moist forest at low and medium altitudes 24
Wooded steppe with abundant Acacia and Commiphora 24
Subdesert steppe 24
Desert 24
Montane evergreen forest 25
Montane communities – undifferentiated 25
Afro-alpine communities 25
Temperate and subtropical evergreen forest. 25
Mangroves 25

II. **Ecological zones of Africa** 26
The use of ecosystems by man. 28
The ecosystem concept. 28
The savanna environment, an example of a man-influenced ecosystem 29
Derived savanna 30
Breakdown of organic material 31
Adaptations of animals to the savanna environment. 31
Biomass and production of consumers. 32
The Guinean zone 33
The Sudanian zone 40
The Sahelian zone 43
The Saharian zone 57
The Mediterranean zone. 60
The Eastern zone 68
The Zambezian zone 77
The Transvalian zone 83
The Basutolian zone. 89
The Kalaharian zone 93

The Karroo-Namaqualian zone 96
The Cape zone 99

III. **Man as an environmental agent** 102
Primitive man's influence on the environment 102
Primitive man's use of fire 102
Primitive man as a cultivator 103
Primitive animal husbandry 103
Implications of the impact of primitive man's influence 104
Modern man's influence on the environment 105
Implications of the human population explosion 107
Effects of fire on the environment 110
Fire and soil conservation 111
Fire as a tool 111
Effects of agriculture on the environment 112
Traditional systems of land use 112
Modern systems of land use 112
The rehabilitation of the Kikuyu lands 113
Variations in the cultivability of land 114
Problems of shifting cultivation 115
Subsistence farming 116
The influence of colonial powers 118
Cash crop production 119
The ground-nut scheme failure in Tanganyika 121
What are range lands? 122
Effects of overgrazing on range lands 122
Effects of degradation of environment on the productivity of wild herbivorous animals 124
Land use and soil erosion 124

IV. **Specific environmental problems** 128
Nomadism and consequences of sedentarization 128
Livestock carrying capacity and land requirements of pastoralists 129
Ecological consequences of sedentarization 130
Marginal lands 130
Arid land management problems 132
The problem of desertification 133
The invasion of Africa by plants and animals 135
Introduced plants 135
Introduced animals 137
The preservation of endangered species 137
The need for preservation of natural vegetation 140
The special need for forest reserves 141
The environmental values of forests 142
The need for national parks or equivalent reserves 143

Animal influences on the grassland environment 144
The role of termites and termitaria 145
The role of the tsetse fly: Africa's boon or bane? 148
The ecology and control of the desert locust 151
The role of the goat 155
Wetlands, estuaries and mangrove swamps 155
The pesticide problem 158
Land tenure problems 160

V. **Problems, needs and potentials in land use** 163
Agriculture 163
Problems 163
The 'Green revolution' 165
Needs and potentials 166
Future trends 168
Range and pasture management 169
Range management on arid lands 169
Prospects and potentials 170
Pasture management 171
Prospects and potentials 171
Animal husbandry 173
Problems 173
Needs and potentials 175
Future trends 177
Forestry 177
Forest production 177
The value of shelterbelts 179
The use of trees in rejuvenating the soil in dry tropical zones 180
Afforestation 180
Future trends and needs 181
Inland fisheries 182
Problems 182
Trends, needs and potentials 185
The fisheries of the Great Lakes of East Africa, a special problem . . . 186
Wildlife 187
Problems 187
Trends, needs and potentials 192
Soil and water conservation 193
Soil conservation and erosion control 193
The control of wind erosion 195
Soil fertility and crop management 196
Green manures and mulches 196
Reclamation of eroded and abandoned land 197
Development of water resources 198
Irrigation developments 199

Water utilization problems 205
Problems in river basin development 208
Prospects in river basin development 212
Food, health and nutrition 213

VI. **Planning for the future** 215
The need for regional planning 215
Ecological considerations in land use planning. 216
Introduction. 216
The need for ecosystem planning 218
Industrial development 220
Environmental quality considerations. 221
Ecological constraints to man's future in Africa. 222
Planning for development: a positive approach to more efficient land use 223

FOREWORD

Africa is not known as one of the more densely populated continents. Yet, the damaging marks of man's activities may be seen there dramatically. Many of Africa's ecological zones are fragile. Large scale soil erosion, resultant cycles of drought and flash floods, downgrading of fauna and flora are well-known to many in general ways, as well as from detailed examination of a few areas. But large parts of Africa remain inaccessible. Very few students of Africa have the opportunity – or the tenacity – to travel over these vast areas or into the hidden corners that lie beyond the well-known routes of Africa. As FAO's Regional Wildlife and National Parks Officer for Africa, ANTOON DE VOS had the opportunity of travelling widely and studying and reporting on the acceleration of man-made changes in much of the continent. As an experienced practitioner of an important and difficult science, ecology, he has made a significant professional contribution with this book. It is our hope that those who read it will be encouraged to carry on the important work and the concern with this subject to which Dr. DE VOS has devoted so much of his knowledge, energy and personal commitments.

B. K. STEENBERG
Assistant Director-General
(Forestry Dept.)
Food and Agriculture Organization
of the United Nations Rome

INTRODUCTION

It is a frightening fact that the quality of the African environment today is deteriorating at an unprecedented and accelerating rate. Many ill-advised, unproductive and destructive practices, such as deforestation, one-crop farming and strip mining, are steadily contributing to an ecological imbalance which has already had catastrophic effects in some areas and will have long-ranging ones in others. If man continues to destroy his environment in this way, he will inevitably produce conditions which will have a strong bearing on himself and even on his ultimate survival.

Unfortunately this detrimental influence of man on his environment is far from levelling off. On the contrary, it seems that the rapid and accelerating population increases foreshadow even more profound and devastating changes. In present day Africa food and farm resources are still recklessly wasted.

Land is lost or misused through bush firing, erosion continues with bad farming practices, labour and time is badly used, storage is inefficient, and processing, transportation and marketing are poorly organized. All this adds up to lost produce and a consequent increase in the number of hungry, ill-fed people on the continent.

The African continent is seriously handicapped by its undeveloped human and physical infrastructure. The acute shortage of qualified personnel, accompanied by an excess of untrained manpower – with over half of agricultural production in the form of subsistence farming – does not bode well for the foreseeable future.

Because man's power is vast in modifying the plant and animal world, very few areas have escaped his impact. A problem resulting from this environmental destruction about which we should be most worried is the general decrease in biotic diversity, that is the oversimplification of ecosystems on a wide scale in favour of domesticated plants and animals and to the detriment of natural, well balanced biotic communities.

In this book an effort has been made to describe the impact man has had on the African environment and to suggest what the immediate and ultimate consequences of his actions are.

It may be appreciated that in view of a scarcity of reliable data from some parts of the continent, documentation is necessarily weak in places.

An effort was made to write this book in such a way that not only the beginning student, but also the educated layman may profit from reading it.

Although most space is dedicated to a description of man's negative influences on his environment, there are indications that with access to modern technology and international assistance many possibilities to counter-balance the downgrading influence of man do exist, which ultimately could aid in an improvement of the environment. It should also be recognised that this impact is not to be thought of as being wholly to the detriment of man's ultimate welfare, because certain man-modified habitats have in fact an equal or greater productivity than the original unmodified habitats.

Our ultimate objective of course should be to try to live in harmony and partnership with nature, with a natural giving and taking. Unfortunately, there is precious little time left to meet this objective.

This book is necessarily biased, because it over-stresses ecological implications and under-stresses socio-economic implications of man's impact on the continent. However, wherever possible, and to the extent of his limited knowledge of this subject, the author has tried to evaluate socio-economic processes. It is realized, of course, that economic development cannot be achieved without disruption of the environment, but such development should aim at minimizing the harmful effects upon the environment itself, upon public health and the welfare of mankind.

I. THE AFRICAN ENVIRONMENT

Because the physical and biotic environment plays such a prominent role in African economic development, a brief outline of the important physical and biotic factors is warranted.

Geology

The geological structure of Africa is relatively simple when compared with the other continents.

Most of Africa consists of a massif of extremely old rocks, formed before the first known terrestrial fossils. These rocks belong to the oldest geological time period, the pre-Cambrian. Here and there on the surface of this underlying block of older rocks occur later rock formations, including sediments deposited in prehistoric lakes and shallow seas which once covered parts of the continent. Strongly folded younger rocks are found only at the margins as for example the Atlas range of the northwest.

During the past sixty million years tropical Africa probably has enjoyed a greater stability of climate and geology than most other large land masses of the world, because it was less affected by the ice ages.

The last uplift of the continent is so comparatively recent that the African rivers have not had time to grade their courses. Many short streams flow swiftly down from the edges of the plateau. A third of the continent has no drainage outlet at all.

Geography

The African continent extends from 37° N to nearly 35° S-approximately one half of it spread between the tropics of Cancer and Capricorn. Because of its position astride the equator, Africa has the highest percentage of any continent in tropical conditions. However, half of the continent is desert, or arid grassland and savannas unsuited to cultivation.

Apart from coastal plains and some river basins, the continent is an immense plateau rising to more than 1000 m above sea level.

Its surface has been subjected to erosion for such a long time that it approximates a peneplain. There are throughout the continent many isolated mountains and some volcanoes which form ecological islands of great interest, because they are occupied by many forms of plants and animals not to be found elsewhere.

Phytogeography

In tropical Africa the vegetation zones have expanded and contracted, have been split and shifted again and again during the Pleistocene Ice Ages. North of the equator were broad vegetation belts similar to those existing today, which moved both in a northerly and southerly direction as a function of warming and cooling trends. The existing vegetation varies from complete desert to luxuriant rain-forest, and in general is dependent upon the amount and seasonal distribution of rainfall. In certain areas, however, altitude exercises an important control.

Some parts of Africa offer problems to the phytogeographer. According to KOECHLIN 1963, it is almost certain that in South Africa several distinctive plant communities exist: that of the Cape region is most specialized; that of the Karroo region, with its very particular xerophytic adaptations which cannot be found in the northern hemisphere nor be linked entirely with the flora of tropical Africa; and that of the Namib desert with its extraordinary plant *Welwitschia Bainesi.*

The floras of high mountains also present problems of integration in a phytogeographic system. The mountains always have many typical orophytes which are distinct from the flora of the surrounding lower areas. These plants can be endemic or, on the contrary, have a wide range.
SCHNELL (1952) describes three mountain districts in West Africa, namely 'le district foutanien', a district covering the central massifs of the Guinean zone and one district in the mountains of the humid zone. HAUMAN (1955) considers the East African mountains exceeding 3800 m to have a high rank on the phytogeographic ladder, because of their very distinctive flora.

Zoogeography

Zoogeographers have divided the world into zoogeographic 'regions'. Africa south of the Sahara has been described as the Ethiopian Region, while the balance is considered part of the Palaearctic Region, which also covers Eurasia. DARLINGTON (1957) has pointed out that the Sahara in fact is in the long run not as important a barrier to the spread of plants and animals as it was once considered to be. Although at present the biota on the northern and southern edges of the Sahara show immense differences, they should still be considered in conjunction.

In Africa south of the Sahara and north of the belt of humid tropical forests the distribution of animals generally occurs in belts parallel to the equator created by the physiognomy of the environment: deserts, steppes and savannas.

The mountains have endemic species of mammals and birds to a greater or lesser extent, depending on their height, and the extent and period of isolation.

The zoogeography of East and North east Africa is the most complex on account of geological disturbances – ruptures, vulcanism, mountain formations, etc.

The continental portion of the Ethiopian region – Africa south of the tropic of Cancer – is subdivided into five zoogeographic provinces:

1. The humid forest province, also called the Guinean zone[1] (Fig. 1).
2. The Guinean savannas, which cover a large part of the Sudanian zone.
3. The savannas and steppes of West Africa, corresponding to the Sahelian zone and parts of the Saharian and Sudanian zones.
4. North east Africa.
5. East and south Africa.

These areas correspond (in general) to the vegetation zones: the distribution of animals is influenced principally by the vegetation, which in turn is determined by rainfall, humidity, temperature and soil.

Fossil records indicate that the area now occupied by the Sahara Desert had a rich fauna, probably similar to that of the present East African savanna. Evidence for this is that throughout the Sahara there are traces of human habitation, fossils and rock engravings, indicating that within the past ten thousand years the area was much more suitable to animal life than at present.

It appears that tropical Africa did not experience the marked fluctuations that occurred in the composition of the fauna and flora of the temperate regions during the past fifty million years and there has consequently been an enormous amount of time for the development of stable and complex biological communities.

Tropical Africa has a greater array of mammals – particularly large ungulates, primates and carnivores – than any other area in the world.

The mammals of tropical Africa are most closely affiliated with those of tropical Asia. Old world monkeys, apes, pangolins, elephants and rhinoceroses are to be found only in Africa and Asia. Endemic African mammals include the giraffe, hippo, and aardvark families, three families of insectivores and six families of rodents.

Very few mammals have a distribution covering the entire continent but among these are found species of shrew-mice, namely a *Suncus* and some species of *Crocidura* (both insectivores), the hare, a porcupine, *Hystrix* and the Libyan cat, *Felis libyca*.

The following mammals are found distributed throughout the Ethiopian region: the hyena, the red mongoose, the elephant, the buffalo and the leopard.

Mammals characteristic of the Saharan area include a species of hedgehog (*Paraechinus deserti*), jerboas (jumping mice), the striped hyaena (*Hyaena hyaena*), the fennec (*Fennecus zerda*) (a desert fox), the addax (*Addax nasomaculatus*) and the barbary sheep (*Ammotragus lervia*).

1 For representative animal species inhabiting these provinces, see descriptions under 'wildlife' in the various ecological zones. The distribution of mammals is stressed more than birds because of their greater economic value and ecological influence.

Fig. 1. The ecological zones of Africa (after DEVRED).

The tropical forest zone is characterized by the following mammals: two species of *Colobus*, the Mona monkey (*Cercopithecus mona*), the chimpanzee (*Pan troglodytes*) and the bongo (*Boocercus euryceros*).

The mammalian and bird fauna of Mediterranean Africa is nearly entirely Palaearctic. Among birds, butterflies and plants, the montane forest species are nearly all different from the lowland species. However, no distinct mammalian montane fauna can be recognized.

The bird species inhabiting the Sahara are neither typically European nor typically Ethiopian.

A comparison of the bird faunas of different vegetation types shows that the moist evergreen forest vegetation type at low altitudes supports the richest variety of bird species, but that these forests rapidly become impoverished with increasing altitude. By contrast, deciduous woodland of various types is generally remarkably rich in species.

Movement of savanna animals between the northern and southern tropics of Africa has been restricted by the presence of the equatorial belt of the Congo forest in the west, the humid Lake Victoria basin, the Kenya Highlands and the arid coastal hinterland of the east. However, about 40 species of birds and some mammals, whose habitat is dry acacia country, have a discontinuous range, inhabiting the Somali area and then after an interval, which includes an area more or less covered by Tanzania and Zambia, reappearing in the dry southwest of the continent. This supports the idea that sometime in the Late Pleistocene during a drier period a corridor of dry acacia country existed across East Africa.

Climate

Because of the large land mass of the continent, there are great contrasts between continental and maritime climates, between the humid tropics and the arid zones, and between high and low altitudes. Africa may be divided roughly into climatic belts following the parallels of latitude; this is particularly the case north of the equator.

The most important single climatic factor in tropical Africa is rainfall, its amount and regularity. The equatorial climate is characterized by heavy and continuous rainfall, generally 175–300 cm per year, and more or less constant temperatures of 26 C or more, day after day. Small seasonal variations in temperature may occur, however. Only a relatively narrow strip (10°N to 15°S), has a typically tropical climate.

In the equatorial and tropical regions rainfall can be detrimental from an agricultural point of view because of the heavy soil leaching which occurs once the monthly precipitation exceeds 20 cm. In those parts of tropical Africa where there are seasonal changes, rain usually falls twice a year. This is caused by a low angle of the sun at the solstices in December and June. With increasing distance from the equator, the two rainy seasons merge into one.

West Africa is in general well-watered by the south west monsoon winds, but elsewhere the great bulk of the annual precipitation falls during the northeast monsoon period of a few months. The amount of rainfall varies greatly from year to year, not only in quantity but also in the date upon which the rain begins.

Africa possesses the highest percentage of arid lands of any continent with the exception of Australia. Only about one third of the continent receives over 100 cm of rain a year, with one third receiving under 25 cm per year. While the dry savanna country experiences heavy seasonal rains, the even drier semi-desert or desert country has irregular and scarce rainfall. In the arid zone of dry heat and desert cold, rainfall is generally too low or unreliable to allow even subsistence cultivation. The region immediately south of the Sahara has great climatic contrasts – nine months of almost complete drought followed by a rainy season.

South of the equatorial belt annual rainfall increases from the west coast, which is semi-arid, towards the east, where the pattern of rainfall is much less simple than in the northern hemisphere. In south central Africa precipitation decreases from north to south.

The coastal areas of North Africa have a Mediterranean climate with a well-defined period of drought in summer and a long period of somewhat irregular rainfall from September to May. The southern extremity of Africa also has a Mediterranean climate with a full winter season in the southeast and rains distributed throughout the year in the south.

Elevation has a profound effect on the climate. The climate of high mountain regions is similar to that of high latitudes, the major difference being that at high latitudes climatic fluctuations are seasonal, whereas at high altitudes on tropical mountains the major fluctuations, at least in temperature, are diurnal.

Long term climatic changes have occurred throughout Africa. These have had a profound influence on the continent's flora and fauna. In fig. 2 are illustrated the broad climatic regions into which the continent can be subdivided.

Soils

The soils of Africa form a patchwork of varied quality. Many of these soils are the result of extremely long development and reflect past rather than present conditions. Sandy soils deficient in important elements preponderate over clay and limestone soils and there are proportionately fewer young, rich, alluvial soils than on any other continent. Among the good to excellent soils are the exceptionally rich alluvial soils of the Nile flood plain and delta, soils of isolated volcanoes, and some of the lowland soils of the Mediterranean region. Some of these soils will sustain permanent crops or repeated cultivation almost indefinitely, or will regain their fertility after a rest period. The best general description of Africa's soils published so far was made by FOURNIER (1963).

Some two thirds of Africa has a soil cover in the sense of a surface layer which supports, or once supported, vegetation of some sort. The rest is stony desert, bare rock, broken rock material, open water, and, at very high altitudes, even ice and snow.

Due to the tropical environment, the majority of the soils occurring in West Africa are highly weathered, poor in humus, and mainly kaolinitic. There are a considerable number of soils, including lateritic soils, which have a hard ironpan developed *in situ* at varying depths of their profile. [illegible] soils, along with black clays, tropical grey clays, saline soils and groundwater

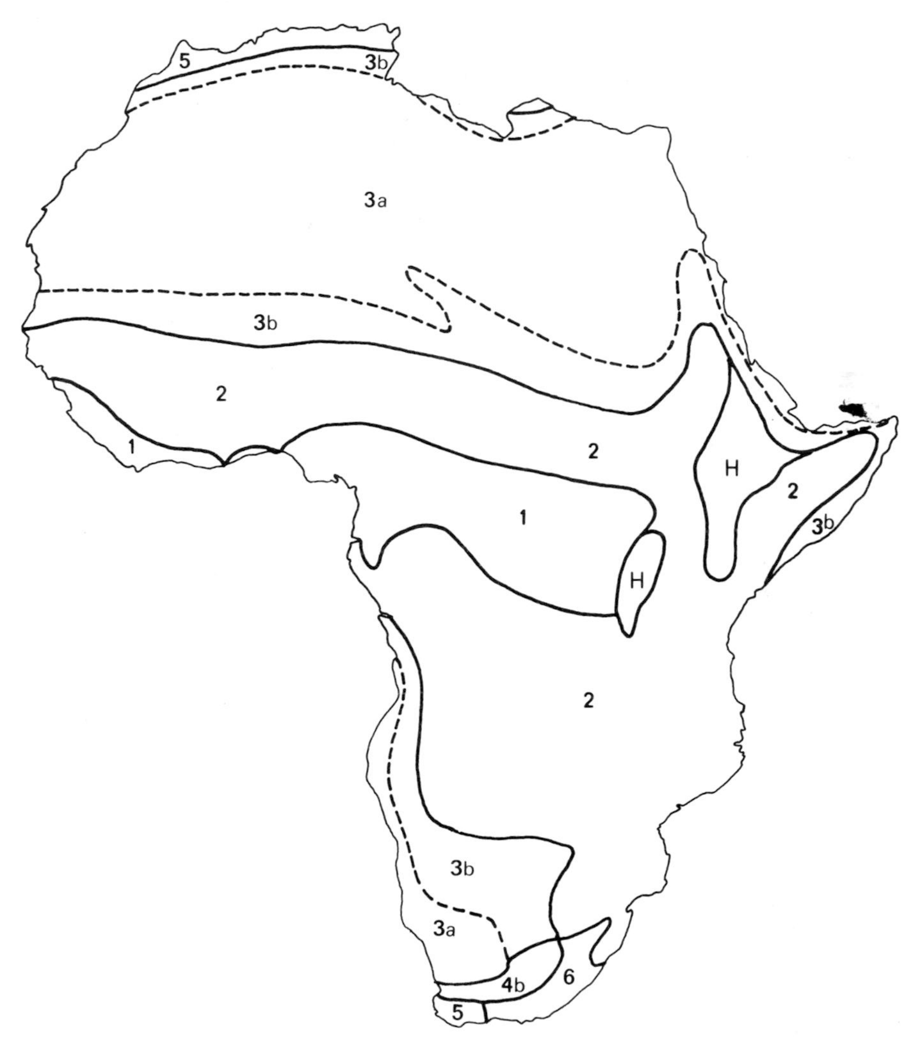

Fig. 2. The climatic regions of Africa (after FINCH & TREWARTHA, Atlas of World Maps, Army Service Force Manual, M 101, 1943). 1. Tropical Rainforest Climate, better called the Equatorial Climate, limited to areas within 10° of the Equator; 2. Tropical Savanna or Summer Rain Climate; 3. Low Latitude Dry Climates, 3a. Desert, 3b. Steppe; 4. Middle Latitude Dry Climates, 4b. Steppe; 5. Mediterranean; 6. Humid Subtropical; H. Humid Tropical (high elavations).

laterites, present many problems in use and management, especially as far as agricultural development is concerned.

The soils of the humid forest environment are generally weakly developed, and relatively poor in nutrients due to leaching. They do have a fair proportion of surface humus, but under conditions of high rainfall become highly acid and possess a very low proportion of exchangeable bases compared with soils receiving moderate rainfall. Humid forest soils are stable as long as the vegetative cover remains undisturbed, but fertility declines rapidly under cultivation. However, under irrigation and careful management they can continue to be productive.

On the whole, the soils associated with the 'Miombo' woodlands of Central Africa are unproductive. Alluvial soils, together with yellow, deep and friable forest and sandy soils occur, but they vary widely over the area.

The richest soils of Africa are black, dark grey or brown. They are known collectively as 'tropical black earths' but have evolved in various ways. They are extensively used for plantation crops like cotton.

The desert soils are produced almost entirely by physical weathering. They contain no humus and are little more than fragmented rocks. Many of these soils are also sandy. The soils of steppe areas, like the desert soils, are also much affected by physical weathering and do not compare favourably in fertility and structure to the grassland soils of the middle latitudes.

It should be remembered that not only centuries of malpractice in land use has caused tremendous deterioration of the soils throughout vast areas of Africa, but that the continent has also experienced a succession of natural erosion cycles because its land surface is so ancient. The effect of both types of erosion on the productivity of the land is similar, but man-made erosion usually causes more rapid down-grading.

Vegetation types

The vegetation of Africa is rich and varied in species. The basis used for the description of the following vegetation types south of the tropic of Cancer is the 'Vegetation Map of Africa' with explanatory notes by KEAY (1959).

WOODLANDS, SAVANNAS AND STEPPES

Most of Africa has a number of rather diverse vegetation types which are generally thought to be closed woodland with little grass as climax vegetation. Rainfall is moderate and severe dry seasons occur. Practically the entire area has a grass cover ranging in height from about 30 cm to 3 or 4 m. Grass fires are liable to occur during dry seasons but most of the trees are to some extent fire-tolerant.

Over wide areas the trees attain 7–25 m in height and form a light canopy over the grass, but with very little shrubby undergrowth present. In moister areas, and where there is protection from fire, shrubby undergrowth may be better developed. The density of tree-cover varies considerably according to edaphic conditions and to the nature and extent of human interference.

One of the important features of the woodlands of tropical Africa is the dominance in the upper story of the genera *Isoberlinia*, *Brachystegia* and *Julbernardia*. In the northern woodlands *Isoberlinia* predominates and in the

Photo 1. Acacia albida and puku in the river bottom lands of the Luangwa Valley, Zambia. Puku are antilopes that frequent visit these river bottom lands, which they graze. Photo: A. DE VOS.

south-eastern areas *Brachystegia* and *Julbernardia* are the most frequently observed genera.

In south eastern Africa – Zambia, Rhodesia and the Transvaal – the dominant tree, tolerant of ill-drained soils is *Colophospernum mopane*. Under optimum conditions, this 'mopane' forms deciduous woodland 15 m or more high, with little or no understory. There are, however, many areas of low scrubby 'mopane', especially in places susceptible to frost.

Moist forest at low and medium altitudes

These tropical forests are evergreen or partly evergreen. Although some trees may be leafless at any one time, the forest as a whole is never completely leafless. In mature forests there are several more or less distinct strata with large trees which may be 40 to 60 m in height. More than 2,000 tree species occur in this type of forest.

A large western block of forest, from Sierra Leone to Ghana, is separated from the main equatorial block by a relatively dry gap in East Ghana, Togo and Dahomey. The main block of this forest stretches from Nigeria in the west to Uganda in the east. A point of interest is the extraordinary and widespread growth of secondary vegetation in this forest as a result of man's activities. Throughout the moist forest regions many freshwater swamps occur.

Wooded steppe with abundant *Acacia* and *Commiphora*

This covers large tracts of land between the Desert or Sub-desert types and Moister Woodland types. It is found along the southern fringe of the Sahara, in the drier parts of East Africa, and in Southwest Africa. The appearance of the vegetation depends on the relative abundance of the trees and shrubs. In some places the trees, mostly species of *Acacia* and *Commiphora* form open to closed woodland or thickets; in other places the trees are widely scattered. Most of the trees are deciduous, fine-leaved, and thorny. The grasses are usually less than 1 m high.

Subdesert Steppe

This covers a relatively narrow belt around the Sahara and in South west Africa which usually merges imperceptibly into Wooded Steppe or Desert types. Low, perennial plants are widely spaced; annuals, including grasses, flourish for a few weeks after rain.

Desert

This type is virtually devoid of vegetation, except for occasional widely scattered plants. The Sahara desert covers an enormous area across northern

Africa. Smaller deserts exist near the Red Sea in Ethiopia, in Northern Kenya and along the coast of Southwest Africa.

Montane Evergreen Forest

This occurs in tropical Africa above about 1,300 m and covers relatively small mountain areas in Ethiopia and South Africa. It is characterized by the comparatively small height of the trees and by an abundance of epiphytic bryophytes.

Montane Communities – undifferentiated

This includes a mixture of montane types, such as evergreen forest, grassland, woodland communities, tree-ferns and thickets of bamboo and occurs in East Africa and Cameroon and covers only small areas. The altitude at which the vegetation occurs varies between 1,300 and 2,000 m.

Afro-Alpine Communities

These occur only above about 3,000 m on the high mountains of central and east Africa and cover very small areas. Representative species include arborescent species of *Senecio* and *Lobelia*, and shrubby species of *Alchemilla* and *Helichrysum*.

Temperate and Subtropical Evergreen Forest

This is restricted to the extreme south. The forests have close floristic affinities with the Montane Evergreen Forests.

Mangroves

Mangrove vegetation occurs in brackish swamps by river estuaries. Mangrove forests develop on mud-flats which are exposed at low tide but are otherwise normally covered by salt or brackish water. The most extensive mangrove forests are in the Niger delta, but they also occur along the East Coast.

For the sake of completeness, mention should be made of a number of other less extensive vegetation types, such as Montane Grasslands, the Forest-Savanna Mosaic, an interphase between the Moist Forest and the Savanna types, the coastal Forest-Savanna Mosaic and the Dry Deciduous Forest. These are not discussed here, because they play a relatively minor role in the overall ecological problems of Africa.

II. ECOLOGICAL ZONES OF AFRICA

I believe that the approach normally used by geographers, namely to describe resource problems confined by national boundaries is not suitable to demonstrate the various ecological problems that in fact exist, because these transcend these boundaries. Instead, an effort has been made to describe these problems on the basis of ecological zones.

The major ecological zones of Africa have been defined by DEVRED (pers. comm.) based on bio-geographic correlations, and he has drawn his ecological sub-divisions on the basis of climatic, soil, phytogeographic and zoogeographic studies of the continent. He paid particular attention to the vegetation in the delineation of ecological zones – following the vegetation map of Africa (KEAY, 1959) – as the interplay of environmental factors expresses itself in the composition and distribution of the vegetation, which in turn controls the distribution of animals. DEVRED's classification of zones largely follows the proposals originally made by LEBRUN (1947). The limits assigned to each ecological zone are merely approximative and transitional, and overlapping areas between zones exist.

The purpose of delineating such zones is to identify large units and to focus attention on general conditions or problems rather than on details. Certain special ecological conditions such as those of the coastal or high mountain areas are not considered in detail in such an approach.

The major ecological zones of Africa are the Guinean, Sudanian, Sahelian, Saharian, Mediterranean, Eastern, Zambezian, Transvalian, Kalaharian, Karroo-Namaqualian, Basutolian and the Cape. They will not be discussed in equivalent depth because some zones are more important from a land use and human survival point of view than others. Particular stress is laid on the ecological and resource management problems which prevail in each zone. Based on differences mainly in the vegetation, DEVRED recognizes several sub-zones in the Guinean, Sudanian and Eastern zones, but these will not be discussed here. The Guinean, Sudanian, Sahelian, Saharian and Mediterranean zones cover broad belts stretched along an east-west axis (fig. 1). Temperature and precipitation are the two main factors responsible for this distribution: they both diminish progressively in a northerly direction with the exception of the Mediterranean zone, where precipitation increases.

In southern Africa conditions responsible for the boundaries of the zones are more complicated. Both temperature and precipitation are greatly influenced by the presence of the cold Benguela current along the west coast and also the warm east coast current. In addition, the great African plateau is highest in the south, where the greatest heights are reached along the south-

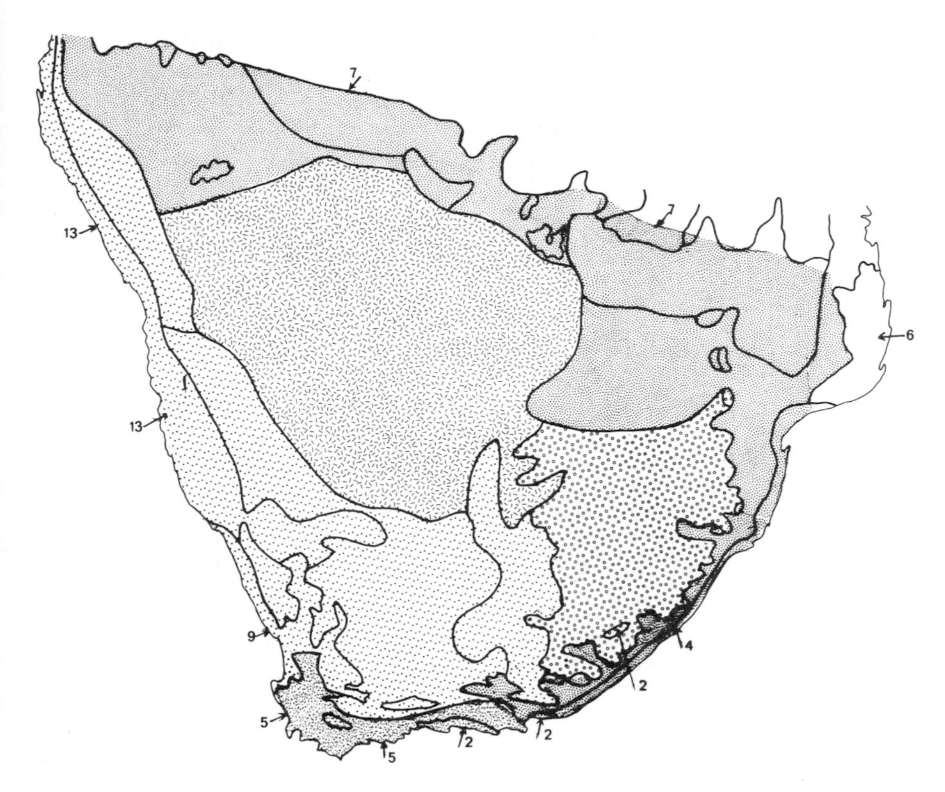

Fig. 4. The ecological zones and natural vegetation of southern Africa (after DEVRED and KEAY); Vegetation types: 1. Montane Evergreen Forest; 2. Temperate and Subtropical Evergreen Forest; 3. Temperate and Subtropical Grassland; 4. Coastal Forest-Savanna Mosaic; 5. Cape Macchia; 6. South-eastern areas with abundant *Brachystegia* and *Julbernardia*; 7. Undifferentiated: relatively dry types; 8. Abundant *Colophospernum mopane;* 9. Karroo Succulent steppe; 10. Subdesert steppe, Karroo shrub and grass; 11. Subdesert steppe, Transitional and mixed Karroo; 12. Subdesert steppe, Tropical types; 13. Desert;

Zones: Transvalian zone; Kalaharian zone; Karroo-Namaqualian zone; Basutolian zone; Cape zone.

east edge and where the surface of the plateau slopes on the whole towards the north and west.

For this reason, the biogeographic correlations appear rather mixed-up. In order to clarify the picture somewhat, a map is included of the natural regions of South Africa, which are, in fact, the ecological sub-divisions of the zones (fig. 4).

The Eastern zone is also quite heterogenous. The plateau is broken by mountain masses which are quite distinct biogeographically. The Lake Victoria area is also quite distinct because of its relatively high rainfall. Finally, the eastern coastal zone has a much higher rainfall than the adjacent areas of the plateau.

It should be emphasized here that the boundaries of the ecological zones, as described, may require adjustments as more data become available. Yet, a comparison with the biotic provinces suggested by DASMANN (1973) which were selected on the basis of similar criteria, indicates that the boundaries of the zones are reasonably accurate. One main difference is that DASMANN recognizes the Ethiopian, Guinean, Central African and South African highlands as seperate provinces, and also describes an East African woodland savanna region. From a land use point-of-view these differences are not very important.

The use of ecosystems by man

During his evolution, man has moved from being a simple component of an ecosystem to becoming a dominant force within it. Through cultivation he has replaced stable ecosystems with associations of his own devising. Man-dominated ecosystems are thus composed of species of plants and animals selected by him for inclusion according to criteria of his own self-interest. Man-made communities are likely to be less stable than were the communities they replaced and are likely to include fewer species. Many of the latter are exotics, introduced for their usefulness. In areas controlled by man vast amounts of energy are removed for food or other products and it is this energy stored by ecosystems that powers most of our human activity today.

The interference of man can modify ecosystems in a dramatic manner in which even the renewal of their resources can be affected. Usually an ecosystem becomes more simplified because certain species are eliminated. Excessive livestock production, deforestation and urbanisation are among some of the many examples of how man can change his environment to his own disadvantage.

The impoverishment of the flora has corresponding consequences on the fauna, more particularly a reduction in the mammals whose lives are directly dependent on plant cover and availability of water. This is also true for man who is restricted in his movements due to the availability or non-availability of water. However, not all interference is disadvantageous, for man can also have a positive influence. He can replace an environment in a natural equilibrium by a more productive one through intensive agricultural practices; he can also reforest to control erosion and he can establish green belts or zones in and around cities for recreational purposes. The human factor should be carefully analyzed because it affects not only the environment but also cultural and economic considerations.

The Ecosystem Concept

Throughout the text frequent reference is made to the existence of natural and man-modified ecosystems. I shall briefly explain the ecosystem concept.

An ecosystem is a more or less closed system in which the mineral and organic resources, solid, liquid and gaseous, are utilized by the plant populations and associated animals in a mutually compatible process. In this process the sun's energy is converted by plants through the medium of photosynthesis to produce chemical energy, forming sugars, starches and cellulose. These plants or the producer organisms, are referred to as primary producers, whilst the animals which may be either herbivorous or carnivorous, are called consumer organisms or secondary producers, in that they then convert the plant material ingested into protein.

The ecosystem is the basic functional unit in ecology, since the properties of the biotic and the abiotic environments influence each other and both are necessary for the maintenance of life as it exists on the earth. All ecosystems require a source of energy, organisms capable of converting this into chemical or food energy and organisms capable of breaking this down into simpler forms which can be re-used.

In an undisturbed area all production eventually enters the decomposer system if there is no emigration or outside dispersal by plants and animals. Such an ecosystem represents a definite level of organization that has been reached by living and mineral matter interacting together. Natural ecosystems favoured by a well distributed rainfall and little change in temperature throughout the year have a much higher productivity than those affected by climatic extremes. It is for this reason that the greatest biotic diversity occurs in the tropics and the least in the deserts and the colder parts of the earth. Natural ecosystems, such as desert communities, may consist of relatively few components. On the other hand, like forest communities, they can be exceedingly complex.

Man has a direct involvement in the the complex set of ecological interactions because he uses one or more products of the ecosystem. These may be biological, as in the case of forest or ranges; physical as in the case of water, or both. Beneficial management involves manipulations of the ecosystem to maximize the returns to man, while exploitation is management that results in the reduction of the productivity of the ecosystem to mankind over a period of time (Van Dyne, 1969).

The Savanna Environment, an example of a man-influenced ecosystem

This particular environment will be described in more detail because it covers vast areas, widely used by man for his husbandry practices and its ecology is rather better known than other vegetation types. An excellent reference for further reading on this subject is 'the ecology of tropical savannas' by Bourlière & Hadley (1970).

The origin of savannas is determined by climatic, edaphic, geomorphic and biotic conditions, or combinations of these. The important climatic requirements are alternating wet and dry seasons. Among the edaphic factors, nutrient status, drainage conditions and geomorphic characteristics are most important.

The savanna environment ranges from humid to arid and there are considerable variations in the vegetation as a result of these climatic differences. The interaction of these factors has produced a complex ecosystem with an extremely high biological productivity. The forest cover, for example, can range from virtually nil to a rather dense woodland. Savannas are characterized by the large number of species of grasses and herbs found at grass level while the upper level is composed of woody plants, ranging from low bushes to tall trees. Many of these trees and shrubs which are so characteristic of the wooded savanna are found to cover old termite mounds, for these afford good growing conditions.

It appears that the existing savanna vegetation is not as 'natural' as was once supposed, and the range of natural savannas has undoubtedly greatly extended under man's influence. In most instances the prevailing climatic conditions could well support a more woody type of vegetation than is at present the case. This is to a great extent a reflection of man's activities, particularly with regard to his annual fires which in almost every dry season sweep across vast areas.

It is now known that this burning has considerably modified large areas of savanna both north and south of the equator. Further modifications arise due to the fact that farmers naturally protect and encourage certain species of trees for their food value such as Shea Butter and *Parkia clappertonia*, or because they consider their presence improves soil fertility, as is the case of *Acacia albida*. Of course, grazing and trampling of the land by domestic stock also leaves its mark. Man's influence on the savanna environment has been in effect for such a long time that even the evolution and range distribution of some wild ungulates has been influenced or altered. It is known for instance, that the spread of some species of antelope has been aided by man, because he created extensive savannas to which these animals could spread. On the savannas of tropical Africa the ungulates are more varied and numerous than on any other continent.

Derived Savanna

It is at its edges that a forest is most vulnerable and it is here that the balance between forest and savanna is most easily upset in favour of the savanna. There is considerable evidence to show that the closed forest zone has shrunk considerably. Areas at the forest edges which at the beginning of this century were still under high forest are now relatively impoverished derived savanna with only a few scattered forest trees remaining.

It appears that forest clearing in itself may modify the climate and result in a reduction of the average rainfall. When the ground is no longer protected by trees, run-off is likely to increase and infiltration therefore decreases. There may, therefore be less water entering the soil even if there is no actual reduction in rainfall. Water tables may thus fall and this combined with the influence of annual burning can make it considerably more difficult for a forest to re-establish itself.

Breakdown of organic material

In a natural area all production eventually enters the decomposer system if there is no emigration or outside dispersal. We have as yet inadequate knowledge of the exact fate of primary production in tropical savannas, such as how much is actually produced and consumed by herbivores and how much passes directly into the decomposer cycle. Certain patterns of organic matter breakdown in some savanna regions are now emerging, however. Thus Hopkins (1966) has compared the decomposition of leaf and woody material and found that whilst micro-organisms are a major influence on the decay of leaf tissue, the breakdown of woody material can be attributed largely to termites. The limited data available suggest a high rate of turnover of materials.

Any tendency towards an accumulation of dead plant materials is usually thwarted by the action of fire. The effects of various burning treatments are not, of course, reflected merely in differences in yield. In many savanna regions in Africa for example, where the community is a fire sub-climax, a whole series of major ecological changes may take place in the absence of fire. The accumulation of dead material not only suppresses new herbaceous growth but leads to a change in the decomposition of the vegetation from open grassland to a scrub condition.

Adaptations of animals to the savanna environment

Animals have several ways of escaping the rigours of dry season conditions. One important way is estivation whereby activity is very much reduced for certain periods of time. A number of adult Amphibia and probably other vertebrates spend at least a part of the dry season in this manner. Another method of avoiding dry season restrictions is by movement, either short or long-range. Thus, Morel (1968) has found that the gallery forest in Senegal acted as a refuge to the savanna fauna during this period. Maximum populations were recorded in this forest towards the end of the dry season, and minimum numbers in the wet season.

More extensive movements are undertaken by many ungulate species. The extent of ungulate movement varies not only from one species to another, but also within the same species in different areas of its distribution. The same species may be migratory or nomadic in one area, yet sedentary in another. The wildebeest for instance, is migratory in the Serengeti plains and is sedentary in the Ngorongoro crater. Both nomadism and migration of wild ungulates may be considered striking adaptations to the seasonality of many savanna areas. In the Serengeti plains, for example, many ungulates move through a vast area of bush and woodlands during the dry season, while during the wet season when fresh grazing and surface water are available on the plains, they move there.

Though some ungulates may cover many hundreds of kilometers a year during their movements, the most extensive seasonal movements of vertebrates are those undertaken by birds. Their seasonal movements may be considered, in part at least, as an adaptation towards the efficient utilization of seasonal food surpluses which the sedentary species are unable to exploit so fully.

Biomass and production of consumers

On the African continent a wide spectrum of wild herbivores is found. In East Africa alone well over 20 species of wild ungulates may live and feed in the same area, within which each species appears to be adapted to a somewhat different set of environmental conditions.

Relatively few studies have yet been made on the energetics of wild ungulate populations in the tropics and the values given for ungulate production on African grasslands are based on data and estimates which are themselves often open to question. At any rate, the productivity for ungulates is very high without showing signs of vegetal and soil deterioration. There is considerably more quantitative information available for these ungulates than for the other consumer groups in tropical savannas, because there has as yet been no comprehensive investigation on the production or role in energy flow and nutrient cycling of the entire ecosystem.

Some of the data recently obtained by a team of workers at Lamto, Ivory Coast, throw an interesting light on factors affecting consumer biomass, or secondary producers in tropical savannas. Biomass data have been recorded throughout the year from some of the major consumer groups in the area such as the above ground arthropods, rodents and insectivores, amphibians and lizards, snakes and birds. Specific data on termite and ant populations are not yet available, though both groups will probably prove of major importance in ecosystem functioning. It was found (Bourlière & Hadley, 1970) that whilst most major consumer groups are present throughout the year, populations are generally reduced during the short, dry season and reach their lowest level following the annual bush fires. Populations tend to increase again throughout the rains, reaching a peak at times which vary from group to group and from one species to another. The groups with highest mobility seem to show the least variation in biomass (and numbers) from month to month.

On considering the total bird biomass one finds that the minimum monthly total, which occurs in December, represents over 60 per cent of that recorded in the most favourable month, July. This is a good example of the effect of rainfall on avian movements and distribution (Bourlière & Hadley, *op. cit.*).

The Guinean zone

Geographic distribution

The Guinean zone covers a vast area stretching from Senegal in the west to Western Uganda in the east, and includes Portuguese Guinea, Guinea, Liberia, Sierra Leone, the southern parts of the Ivory Coast, Ghana, Togo, Dahomey, Nigeria, Cameroon, and the Central African Republic, Zaïre, Congo (Brazzaville), Equatorial Guinea, Gabon, Ruanda and Burundi.

Climate

The climatic boundaries of this zone are set at the limits of more or less four months of dry season. The main climatic features are the relatively even distribution of rainfall throughout the year and the lack of major fluctuations in temperature.

Precipitation, exceeding 1000 mm per year, is either distributed evenly all year round the equatorial regions or in the sub-equatorial regions occurs in long rainy seasons interrupted by dry ones. The rainfall has two maxima and two minima per year, either simply marking the existence of more or less rainy seasons, or else delineating two extremely dry and two rainy seasons. The mean annual temperature is around 26 °C and there is virtually no contrast between the seasons.

The relative monotony of the climate hides variations due to the geographic location or topography but just the same, very distinct Guinean subclimates can be recognized. For example, the dessicating influence of the Harmatan (a dry wind originating in the Sahara) is felt much more strongly in Nigeria than in Ruanda. In the latter area monsoon influences are felt.

Vegetation

According to the vegetation map of Africa this zone includes moist forest at low and medium altitudes, bordered to the north and south by forest – savanna mosaic. The typical ecosystem of the Guinean zone is the tropical rain forest. Under natural conditions such a forest is a very stable ecosystem which would be dominant throughout the zone were it not for the influence of man.

The Guinean vegetation exhibits rather significant variation. This is particularly apparent between extreme western and eastern parts of the zone.

Most of the original vegetation of the Guinean zone was either vast jungles of dense evergreen rain forest or of moist semi-deciduous stands. As many as 500 tree species have been described from the humid forest in Ghana. Around the periphery of the main forested area semi-deciduous gallery forests or isolated forest stands are to be found among the savanna lands.

Generally, the savanna lands result from man's encroachment on the forest. The plant cover is now dominated largely by grassland, forest-substitute bush, clearings, forest re-growth, or secondary stands that have developed since clearance.

As man's interference with the forest continues, these stands are spreading at the expense of forests. The approximate historical limits of the Guinean forests are not certain, but it is likely that more than half of the original forest area has been sacrificed to wooded savanna and agricultural lands. Considering the relatively brief time span – a few centuries – in which this occurred, this is a very drastic environmental change.

The oil palm, (*Elaeis guineensis*), is found growing throughout and is the most typical forest crop. It is of great economic importance, and although most oil comes from wild trees, huge plantations for the production of oil exist in Zaïre and Nigeria.

Several indigenous plant species are now in danger of extinction due to replacement by introduced varieties of crops. Amongst the disappearing species are African rice (*Oryza glaberrima*), bambara nut, and yams.

This zone is rich in fruits, nuts and oil plants in an incipient stage of domestication. Perhaps of great potential in this area are medicinal plants, some of which are at present important forest products but which might later be developed into formal crops. Other wild species such as *Citropsis* are of potential use for their value as rootstock for citrus, and another important group, consisting mostly of wild species, are the oily and fatty seeds such as those of the Shea butter tree which are consumed locally in great quantities. The zone also contains some native spices such as the *Melequeta* and 'grains of paradise' and these along with stimulants such as kola nuts may develop into secondary crops in a diversified agriculture.

Food from the forests

Many foods which provide energy, protein, vitamins and minerals to the diets consumed by the people are derived from forest crops not primarily cultivated for food. Their importance is often ignored and frequently discounted by nutritionists in their evaluations of available food. Palm wine produced from the palm tree *Raphia* sp. and also from the oil palm is a source of vitamin B contained in the sediment of the wine. The latter tree provides the major source of vitamin A to West Africans in the form of red oil which contributes more than one-tenth of their energy intake.

Photo 2. The present rapid increase of populations is most apparent in developing countries, many of which have an average population growth rate of 2.5–3.0% annually. More mouths to feed in West Africa mean that more land must be brought under cultivations. This involves much back-breaking work when only primitive handtools are used. Africans clearing a piece of land for cultivation. Photo: FAO.

In the fresh water swamp areas this tree exceeds all others in its importance and provides the indigenous people with plentiful supplies of protein from the kernels and carotene from the pericarp. The surplus is exported as red palm oil and kernel oil. The oil palm also harbours an important delicacy – the larvae of the rhinoceros beetle (*Rynchophorus phoenicas*), which is a significant source of protein. In these swamp areas many other trees such as the oil bean (*Pentaclethea macrophylla*), and the dika nut (*Irvingia gabonensis*), provide additional important sources of protein in their nuts or kernels.

The miraculous berry (*Synsepalum dulcificum*) is used in the production of soft drinks. Taken just before an acidic food, such as lemon, this berry adds a delightfully sweet flavour to the otherwise sour taste.

Effects of human influence

Human activities have been marked by a progressive diversification in land use. Until the start of this century human activities, with the exception of fire, caused relatively little disturbance to the forested environment. It ceded areas to settled agriculture and furnished the required products with little danger to its stability. In more recent times, however, there is a growing realization that accelerating changes are taking place, particularly in the densely populated areas. Ecologically speaking, these changes are largely negative in character, causing a degradation of and a reduction in the stability of the ecosystem. Some permanent crops, however, such as oil palms, cacao, rubber and coffee do not share those distressing effects on the productivity of the land because the soil remains well protected and the amount of nutrients removed from the ecosystem in harvesting is not large.

Forests

If degradation of the forest does no more than cause a change in species composition, whilst still preserving the structure of a closed forest, it may still have economically serious implications though not necessarily environmental ones.

Considerable and lasting damage has already been done to the Guinean forests. Vast areas have been clear-cut for agricultural and grazing purposes and contrary to popular belief there now is little tropical rainforest left untouched by man's hand. Many of the forests that at first sight appear

---

Photo 3. The farmers of Togo are industrious people whose traditional practices demonstrate an inherent knowledge of good land and water usage. To control erosion they build plots of earth mounds on which they plant either cassava or yams. The yam tubers than have loose soil in which to grow, while the excessive rainfall which occurs in this area will not flood the crop. Maize is planted in between the earth mounds. Farmers are working in a plantation of yams. Photo: FAO.

'virgin' are in fact utilized by primitive man such as the pigmies for the hunting of animals and gathering of plant products. These people do not affect the composition of the forest much, however. The system of bush fallowing, increasing population pressure, and the widespread use of fire has lead to the disappearance of the forest and is endangering the supply of forest products. For example, the big area that appears on maps as forest in southern Nigeria and Ghana is now mostly under crops of one sort or another. Of the surviving forest little, if any, seems to be in a truely undisturbed condition.

Although the forests of this zone have been seriously over-exploited, they remain the most important timber resource of the continent and considerable quantities of lumber are still being exported.

More appalling is the fact that there are thousands of square miles of now unproductive grassland on which the forest has not the slightest chance of regeneration. In addition to this unproductive grassland, large areas of woody, shrubby vegetation of very little use to either man or beast stretch over the landscape.

Agriculture

Felled forest and forest edge provide excellent conditions for the cultivation of staples such as bananas, cassava, taro, yams and sweet potatoes.

The cassava (*Manihot utilissima*) is the staple food for the greater part of the population. It was brought to Africa by the Portuguese and due to its many advantageous qualities it spread very fast, pushing aside such indigenous cultivated plants as taro (*Colocasia antiquorum*). The yam (*Dioscorea alata*) has also spread, particularly in the rain forest regions of the countries on the coast of Guinea. Cassava is commonly grown on the poorer forest soils, on exhausted land, and in abandoned cocoa plantations.

In the forest area important crops include coconuts, mangoes (*Mangifera indica*), cattley guava (*Dacryodes guava*), *Psidium guajava*, papaya (*Carica papaya*), and citrus fruits. Para rubber trees (*Hevea brasiliensis*) have been planted in rather localized areas. Coffee and cacoa have become widely established, either as plantation crops or as private farm products. These latter two crops have been for several centuries and still are, to a lesser extent, important exports.

The major food crops include cereals, principally maize, rice and sorghum; protein rich grain legumes and oil seed crops of which cowpeas and soya beans are amongst the most promising from the economic point of view; root crops and a number of vegetable crops. Groundnuts are grown partly for export.

Cotton is grown on a large scale in the drier parts and sugar cane in the more humid parts of the zone. Congo jute (*Urena*) is grown locally as a plantation crop.

Wildlife

Despite heavy hunting pressure and widespread habitat destruction there are still substantial wildlife populations in this zone. Typical species are the western gorilla (*Gorilla g. gorilla*), the olive colobus (*Colobus verus*), Jentink's duiker (*Cephalophus jentinki*), black duiker (*Cephalophus niger*), banded duiker (*Cephalophus zebra*), Royal antelope (*Neotragus pygmaeus*) and the chevrotain (*Hyemoschus aquaticus*). The okapi (*Okapia johnstoni*) is restricted to a relatively small area in Zaïre. Elephant and dwarf forest Buffalo are widely distributed as are the hippopotamus, the giant forest hog (*Hylochoerus meinartzhageni*) and the western bush pig (*Potamochoerus porcus*).

The pigmy hippopotamus is very rare. Unfortunately, as a result of increasing human population and land use pressures, wildlife populations and habitats are, in general, still decreasing. However, not all wildlife species show a steady decline. In fact an area cleared of most of its trees and planted with agricultural crops may often support larger wildlife populations than before because of the greater variety and availability of food. Several species of duikers, cane rats and many species of birds are, for example, often more abundant in such cases.

In many of the more densely populated parts of the zone, the larger herbivores, such as elephants, hippos and several medium-sized antelopes, such as the kob and waterbuck, have been virtually eliminated. The same is true for the larger predators such as the lion, leopard and hyena, although the latter species is locally common.

The bird fauna is rich in number of species. Guineafowl, francolins, spurfowl and other species commonly utilized by hunters are widely distributed. Less pronounced changes have taken place in the avifauna than among the mammals because they were hunted less frequently, but many of the true denizens of the tropical forest, like hornbills and parrots have either diminished in numbers or completely disappeared.

There are many large and medium sized rivers and lakes in this zone and most of these waters are well supplied with nutrients. Thus substantial fish populations exist, and these are regularly exploited.

Rangelands

The rangelands were originally created by man through the cutting of forest and by large scale, repetitous burning. In contrast to the rangelands of the more arid parts of Africa, these rangelands are not generally very productive, being largely covered by unpalatable grasses. They are poorly stocked with cattle, sheep or goats because the majority of the Africans who inhabit this humid, tropical zone are generally not livestock producers, and the livestock that does exist is inferior in terms of reproduction and meat production potential. In tsetse-free areas zebu cattle are kept and breeds tolerant to trypanosomiasis occur elsewhere.

Panicum gracilis and *Stiloxanthus*, both grasses, have been sown in order to improve the pastures and have been very successful over a wide range. The latter species remains green in the dry season and can be used to advantage on poor sandy soils. *Desmodium contortum* can also take the place of *Stiloxanthus* when used at higher altitudes.

Prospects

It seems that the process of forest destruction will go on unabated, with the exception of a few remnant stands preserved in forest reserves and national parks. With the increasing demand for forest products more emphasis will be given to the establishment of plantations and regeneration of still existing forests. This is already in progress for example, in Nigeria where extensive plantations of teak have been established.

The production of swamp rice is gradually expanding from Guinea and Sierra Leone to other countries (FAO, 1968). The higher potential yields of this type of rice make it more promising than upland rice and therefore its cultivation should be enhanced wherever feasible without causing undue damage to available fish and wildlife resources.

Livestock production is a real possibility for the future of this zone and I believe that increase in income and improved stability can be achieved through improved mixed farming practices. This of course, requires considerable improvements in pasture management. Pork production should also be concentrated on as the limited efforts that have been made so far indicate that these animals can materially assist in supplementing the protein supply without too much improvement in farming practices or financial outlay.

The Sudanian zone

Geographic distribution

The Sudanian zone is an area of extensive semi-arid lands south of the Sahelian zone, stretching across Africa from the Atlantic coast to the mountains of Ethiopia. This belt includes most of Senegal, the southern parts of Mali, Upper Volta, Niger, Chad and the Sudan, and the northern parts of the Central African Republic, Ivory Coast, Ghana, Togo, Dahomey, Nigeria and Cameroon.

Climate

The main climatic characteristic is a single rainy season which occurs during the northern summer. Rainfall varies between 600–1,250 mm. Nowhere else in the world are such uniform climatic gradients found over such a large area as exist in this vast savanna belt. These gradients are aligned from north to south, and thus similar conditions are found along similar parallels of latitude. Temperature differences between day and night are considerable and increase near the Sahelian zone. The growing season ranges from 70 days in the north to 250 days in the south. During the dry season, which last 5–9 months, the Harmattan, a dry wind, blows from the north east, while during the wet season rains are brought by a moist south westerly monsoon. The temperature is high throughout the year; The mean monthly temperature rarely dips below 21 °C. There is little change in temperature during the year, and the drop in temperature brought about by the summer rains is usually greater than that caused by the more oblique sun rays in winter.

Vegetation

The vegetation consists mainly of open woodland savanna, containing mainly deciduous trees. The number of tree species is much more restricted than in the Guinean zone. In the southern, more dense and relatively moist woodlands *Isoberlinia* species are the dominant trees and *Hyparrhenia* is the dominant grass, while in the northern relatively dry savannas *Andropogon* is the dominant grass species.

Effects of human influence

Rapid increases in human population densities in this zone and corresponding increasing pressures on the land have resulted in much land degradation, the signs of which can be readily observed anywhere within the region. Dust storms have become a more frequent phenomenon and widespread desiccation of the land has taken place.

Agriculture

In the south of this zone cropping is important, and normally a fallow system is used. Millet (*Pennisetum* spp.) and sorghum (*Sorghum vulgare*) are not only the staple foods in this zone, they are also used in the preparation of alcoholic, such as millet beer, and non-alcoholic beverages. Millet is grown in the drier parts of the zone while sorghum and cassava are found in the areas with higher rainfall. Groundnuts (*Arachis hypogaea*) are grown in most of the countries of the region and form an important base of the local diet, providing a source of protein and vitamin B. Rice is widely grown in riverine areas. Other crops widely grown for human consumption are grain legumes, including cowpeas (*Vigna sinensis*), beans (*Phaseolus* spp.), Cambarra groundnuts (*Voandzeia subterrenea*) and a group of miscellaneous species such as bananas, which are grown in restricted quantities.

Animal husbandry

Livestock represents a major economic asset in this zone. Practically all the feed comes from unimproved grasslands or stubble fields and fallows which are grazed after harvesting. In the western part of this zone considerable numbers of animals are maintained by the settled farmers, in addition to the migratory herds. The value of organic matter is recognized, and manure is carried to the fields. Cattle are not only the native cattlemans' main possession but also his very means of sustenance provided chiefly from milk. Heavy livestock losses are incurred during times of drought. When at the start of the dry season surface water disappears, graziers move their herds to the more humid areas further south. With the first rains, however, cultivation displaces the cattle and they return to the drier areas in the north. These movements have repercussions on the environment due to trampling by the moving herds and also on the condition of man and beast. The effect on the latter is significant, for the travelling distances involved cause severe loss of liveweight and the rate of maturity is consequently low.

Sudanian livestock includes cattle, goats, sheep and poultry and restricted numbers of donkeys, camels and dromedaries. There is a basic need for an increase in animal production which can be brought about by the provision of better pastures, principally on irrigated lands, along with such necessities as supplementary feed and adequate water supplies during the dry season.

Large areas of eroded, wasted land exist throughout this zone and much of this is due to constant overgrazing. This has led to the disappearance of the more palatable perennial grasses and thence to an increase in bush and weed growth. Desert conditions continue to spread in most marginal areas and if overgrazing continues it is likely that ever more land will become unproductive.

Forestry

Forestry is of much less importance here than in the Guinean zone. While in primitive times no doubt good quality forests occurred in the southern part of the zone, existing woodlands have been badly depleted for building materials, fuel and fodder for livestock. It is the habit of nomadic people to lop off branches of trees during the dry season for the use of their flocks of goats and sheep, but this ultimately results in reduced vigour and eventual death of many trees. As little suitable timber remains, reafforestation efforts are necessary. During recent decades many tree species have been introduced, such as several species of *Eucalyptus* and the Neem tree (*Azadirachta Indica*). The latter tree grows particularly well and is used for the production of charcoal. In addition, it improves the soil due to nitrogen fixing bacteria contained in nodules in the roots.

Prospects

Programmes of soil and water conservation need to be improved if agricultural land is to be kept permanently productive and water resources retained.

The potential contribution of livestock to the economic development of the countries concerned has not been realized. As the overall productivity of stock in the area is low, major concern of the countries within the zone should be to improve the exploitation of pastures. Work will continue to create a network of water holes to provide livestock with sufficient, well-distributed water and thus avoid the necessity of long migrations.

Future improvements must be geared to existing ecological, sociological and economic conditions and will have to be achieved through the adoption of improved feeding methods, management and breeding of livestock. Once disease losses are controlled, a regulated take of slaughterable livestock must be established in order to maintain stable population numbers.

An urgent need is the reduction of the present rate of land degradation but the difficulties of overcoming the problem are considerable due to the present nomadic systems in operation. Attempts are now being made to demarcate grazing lands and water reserves are being developed to extend the normal grazing periods longer into the dry season. Supplementary feed has been issued in certain areas during the dry season.

Free-ranging poultry have been introduced in nearly all settlements and provide considerable amounts of protein to their owners. Modern and large-scale broiler and egg units are slowly developing.

The area under cereals is increasing. The need for the extension of edible grain legumes cultivation and for the improvement of productivity is enormous. Their importance is based not only on the fact that they are one of the cheapest sources of protein for human consumption but that they also improve the productivity of the land, resulting in higher yields of the crops that follow them. The present critical situation in the production and consumption of edible grain legumes is due to low yields caused by lack of adapted high yielding varieties, poor cultivation practices and by serious insect pests and disease problems. With greater concentration on higher yielding varieties and their correct cultivation, much could be achieved.

This zone is particularly suited to a large development of irrigation and flood control works during the years to come.

The Sahelian zone

Geographic distribution

This zone covers a rather narrow belt of very arid country, 200–250 km wide, occurring between the Sahara desert and the northern edge of the Sudanian zone and situated roughly between latitudes 14 °N and 20 °N, and longitudes 10 °E and 23 °E. It includes parts of southern Mauretania, northern Senegal, northern Mali, Upper Volta, Niger, Chad and the southern Sudan. Although most of this zone is quite flat, it is interrupted by four mountain massifs.

Climate

The climate is characterised by erratic, scattered rainfalls occurring during the northern summer. Total rainfall amounts to less than 500 mm per annum, but seldom averages more than 250 mm in the northern areas, As a consequence the growing season is generally very short, being dominated by rain showers. Dry spells are frequent. Winter temperatures are reduced to a mean daily maximum of 21 °C from 30 °C in summer.

Vegetation

The vegetation of this zone is classified in the vegetation map of Africa as subdesert steppe – tropical type and is dominated by the genera *Combretum* and *Acacia*.

The vegetative resources of this zone depend largely on two preponderant climatic features: rainfall and evaporation. Plants also depend to a great extent on the texture of the soil, as this determines the water retention capacity. This milieu expresses itself by a special formation and evolution of the soil and a special geomorphology, both affected by extreme variations in daily temperatures, wind erosion, etc. The dynamic equilibrium which exists between the plants and their physical environment is hence very precarious. The hardy perennial plants that do exist are widely spread covering 10 to 80 per cent of the available land surface. Burdened by an already arid climate it doesn't take too long before desertification sets in if fire or overgrazing get a foothold.

Many plants have special physiological adaptations to minimize water loss through transpiration, such as spines, reduced leaves, etc. Range vegetation consists of a low layer of annuals and higher layers of perennials, shrubs and trees. During the dry season the vegetation is dormant with the exception of the flood plains of rivers and swamps. Flood plain pastures occur along the Senegal and Niger rivers and lake Chad. These are accessible for grazing during the dry season.

Photo 4. Five years of subnormal rainfall brought severe drought and tragedy to Upper Volta in the Sahelian zone in 1973. During most of the year, the edge of the Sahara desert in eastern Niger is a parched, arid region. It is here that several nomadic tribes roam on their never-ending search for water for themselves and their cattle; it is upon the latter that their livelihood depends. Portrait of a Foulani nomadic cattleman among his herd of zebu cattle. Photo: FAO.

The ecosystems of this zone reflect the great severity of the physical factors involved. Thus there is weak primary production, a reduction in animal populations and a concentration of man and his stock around the watering places, marshy areas or regions inundated by water. Primary production can, however, be high in certain natural pastures allowing good secondary production as is the case of the savannas.

Because of the recurrent droughts and consistent over-use of both herbaceous and woody vegetation, environmental conditions are generally poor and deteriorating, and the process of desertification is going on at full force. The nearer to the desert, the more susceptible an area is to mismanagement.

Human and animal health measures have contributed to an increase in the populations of both. The lack of any range management policy has led to overgrazing, particularly around watering holes. Water development programmes – the drilling of deep bore-holes, for instance – have been established without proper regard for the surrounding land.

Increasing population coupled with increasing demand for food and firewood, widespread use of firearms and social measures, such as the tentative fixation of nomadic tribes by various governments, have all resulted in abuse of the scarce natural resources and have disrupted the traditional land use pattern. An increasing conflict exists between nomads and cultivators. The latter are more and more inclined to push their endeavours to the north. This results in increasing difficulties for the nomads to have access to rivers and marshes as the cultivators grow millet and sorghum on land that has become free of inundation.

AGRICULTURE

Agriculture is possible only in oases and temporarily inundated lands. On the climatically more favorable parts of the zone crop cultivation is practiced. When rainfall is below 400 mm, cropping is possible only in areas which have been flooded during the wet season. Sorghum is usually grown but rice is also found. Above 400 mm of rainfall crops of pearl millet and also sorghum occur. The most important hazards and limitations to crop production include, in addition to the scarce and erratic rainfall, the *Quelea* bird and the locust which do extensive damage to crops. Where irrigation is possible it is practiced but usually with somewhat primitive methods. The opening up of land for cereal culture marks the destruction of the last remaining marshy area's in this zone

---

Photo 5. Five years of relentless drought reduced the six nations of the Sahelian zone to the most tragic situation in their history. Dying animals such as this donkey were a common sight during the drought of 1973 throughout the sub-Sahara. This burnt-out and parched land in Niger once provided nomadic herdsmen with relatively decent pasturage. Photo: FAO.

which are now threatened by conversion into rice fields. Irrigation is widely practised and several irrigation projects under international and bi-lateral assistance schemes have at least been partly successful in the production of more food and fodder.

Forest Vegetation

Tree cover occurs over most of the zone and trees increase in frequency as one approaches the Sudanian zone. Cover is open, that is to say that there are large distances between individual trees and shrubs. The trees are often spiny with small or frail leaves, stunted and parasol-shaped. Typifying species include *Acacia senegal*, *Commiphora* spp., *Bauhinia rufescens* and *Lannea humilis*. Other *Acacia* species include *A. seyal*, *A. mellifera*, *A. ataxacantha* and *A. raddiana*, and other tree species are *Combretum* and *Pterocarpus*.

Adapted to coping with a long, dry period, these trees slow down or stop their vegetative activity during the hot season. On the other hand, they are equally capable of taking maximum advantage of slight rainfalls.

The major use made of trees is to lop off branches to feed goats and sheep, particularly during the dry season. *Acacia seyal* is the most favoured species for this purpose. Trees and shrubs are also used for firewood, and around towns and villages much land is denuded of trees as a result. Deforestation has had as one consequence the increase of a much larger number of xerophyle insects, such as grasshoppers, which in turn have caused much damage to the vegetation. Gum-arabic is produced from *Acacia senegal* and *A. seyal*, particularly in the Sudan, which produces 85 per cent of the world output (Fishwick, 1970).

Tree cover is essential in this zone for erosion control and is helpful in soil regeneration. This has long been recognized and for this reason efforts have been made in reafforestation and the planting of shelter belts, using mainly exotic species such as *Eucalyptus* spp. and Australian acacias. Effective reforestation can also be accomplished with *Tamarix*, *Acacia senegal* and *A. laeta*, which are indigenous.

Wherever natural tree cover remains it should be carefully protected. Particular attention should be paid to remnant *Juniperus* stands in the Sudan above 1250 m, because it is the tendency of this species to disappear. *Podocarpus* also faces danger of elimination, as it is being over-utilized for the production of roofing poles.

Photo 6. Emaciated cattle in Upper Volta manage to reach a water hole. But with pastureland ravaged by drought, their chances of survival are poor. FAO photo taken during the drought of 1973.

Nomadism

By far the largest group of people inhabiting this zone are the nomads whose fixed points of life are the rare sources of water, wells or temporary wetlands. Naturally an extreme concentration of life exists in the oases or areas which are irrigated. Most nomads have relations with the inhabitants of oases under which they are allowed to make at least some use of the products.

Livestock production

Livestock is produced either by the nomadic system, in which the nomads move around in search of pastures on which rainfall has recently fallen or by the transhumant system, by which they move southward during winter and northward during summer to take best advantage of growing conditions. The nomadic system is practiced where rainfall is less than 150 mm.

Livestock consists mainly of camels in the northern part, and zebu cattle, goats and sheep in the southern part.

The availability of water is a severe limiting factor. Cattle are more restricted in their movements than sheep and goats because they are poorer walkers and big drinkers. The extent of the range used from a watering point should vary with the time of the year. At the start of the dry season it should not exceed 10 km for cattle and sheep if the animals are not to lose too much energy finding water. By the end of the dry season it should not exceed 5 km. Productivity is limited by the low carrying capacity of grazing lands. As a result of rising human populations and correspondingly increasing livestock populations, the range has deteriorated in many parts of the zone (see further p. 123).

Often times the range is so depleted that cattle populations can no longer be sustained, in which case goats and camels take over.

Rangeland

Productivity is limited by the low carrying capacity of grazing lands. The rangeland vegetation is composed mostly of annuals as the useful perennials have frequently been eliminated by overgrazing. Production is therefore so limited and low in quantity that it has been estimated that branches lopped off seasonally by shepherds from trees contribute more to the diet of the livestock than the grass cover. Stubblefields are grazed after harvest by transhumant animals, but their contribution to the production of food is not very large.

Efforts have been made to improve the range, particularly by the plantation of fodder producing trees such as *Acacia albida*, Australian acacias, and *Atriplex* species. These experiments have been generally successful and should therefore be extended on a much larger scale.

Photo 7. Five years of subnormal rainfall has brought severe drought and tragedy to Upper Volta. Destitute nomads from Mali came to northern Upper Volta in search of pasture but found a wasteland. The carcass of a cow lies in the foreground. Photo: FAO.

Photo 8. Carcasses of animals lie around a camp of nomads in northern Upper Volta. Photo: FAO.

Photo 9. Subnormal rainfall has brought severe drought and tragedy to Upper Volta. This picture, taken in June 1973, shows Peul nomads who have lost all their cattle and sit forlornly in a camp near Markoye. Photo: FAO.

Wildlife

Until recently this zone was inhabited by a spectacular fauna of wild herbivores, including giraffes, roan (*Hippotragus equinus*), and several other species of antelopes. Elephants, buffaloes and warthogs (*Phacochoerus aethiopicus*) were restricted to those areas where they had good access to water. Wildlife has suffered greatly from habitat deterioration and overhunting. Were it not for the existence of some national parks and game reserves, some larger species such as elephants would have already been wiped out. The addax and oryx are large antelopes which live for extended periods in the absence of water and offer an important source of protein in those parts of the zone which are too arid for livestock production. Unfortunately, hunting from vehicles has caused a serious decline among these species. Ostriches were once widespread but are much persecuted for their skin, which makes excellent leather, and their eggs. They are now much restricted in range. Spotted cats like cheetahs and leopards have much declined in numbers.

During the northern winter there are considerable concentrations of wildfowl on the rivers, in estuaries and on temporarily inundated lands. As a result of the recent droughts wildfowl populations have been drastically reduced, and this is equally true for the other fauna.

Prospects

Droughts, as well as serious livestock losses and crop failures, have always followed a recurring pattern and will continue to do so. What has been experienced in the crisis of 1973 is the cumulative combined effect of several exceptional dry years and land degradation caused by man. This has highlighted the vulnerability of the present production systems. There is no doubt that as long as the degradation of the land continues similar crises will reoccur, each time probably with greater severity than before. The recent succession of dry years has brought to the fore the cost of neglecting long-term planning and the need for proper attention to limiting factors caused by ecological conditions in the development of the zone.

Crop production should be improved and mixed farming promoted. This can be done through the development of fodder crops, greater use of animals for traction, the use of small-scale irrigation, the incentives to small scale fattening operations, etc. Such a programme is certainly dependent on the improvement of cereal yields.

Overgrazing, the most serious problem of this zone, can only be effectively solved on a long-term basis, by better education of pastoralists in range management, improvements in land tenure and, last but not least, control of stocking rates of livestock. Under good management considerable surpluses of livestock can be achieved. Some of the West African countries bordering the Sahara should be able to export large quantities of animal products to their neighbours further south who with their problems of large populations

Photo 10. The appalling effect of hunger can be seen in the faces and bodies of these destitute nomads from Mali. Photo: FAO.

and relatively poor livestock resources create an excellent market.

The lack of integration between the livestock and crop belt hampers both a possible de-stocking of range areas and the improvement of the soil fertility of the crop lands. There is a growing deficit of cereals, which has to be met by imports, consequently, there are no food reserves which can be utilized in case of emergency.

More irrigation schemes using surface water will be developed along the major rivers flowing from rainy areas. The Nile and the Niger are of primary importance, and the Logone and Chari secondary in this respect. It is essential that in the future a careful assessment is made of the impact of such schemes on the ecology of the environment and on the sociology of the local inhabitants, so that mistakes in development and management can be minimized.

Control of *Quelea* birds and insect pests should continue unabated, but in control projects more attention must be given to the ecological problems of control, and more particularly to integrated pest control.

It is urgent to work out a rational management for the ligneous resources of this zone. Reafforestation and the establishment of shelterbelts should be stepped up and integrated with efforts to stop desertification and to aid soil and water conservation. Species to be selected should not only include indigenous species of *Acacia*, but also various introduced species, including *Eucalyptus* and *Acacia* species from Australia. Reafforestation should be diversified so as to comprise areas protected from grazing, fixation of dunes and the planting of trees on arable land.

Before addax and oryx populations are wiped out altogether, a concerted effort should be made not only to safeguard the future of these drought-adaptable species, but also to increase their numbers so that they will again serve as a food supply to the local inhabitants.

The Saharian zone

Geographic distribution

It is somewhat difficult to describe the geographic limits of the Sahara desert, as it continues to spread yearly. However, in the west it reaches the Atlantic Ocean in Mauritania and Spanish Morocco, and in the north west is separated from the coastal areas by the Atlas mountains. It reaches the Mediterranean Sea in many places in Libya and Egypt, where it also stretches as far south and east as the Red Sea, but is known as the Nubian desert east of the Nile. The northern parts of Mauretania, Mali, Niger, Chad, the southern parts of

---

Photo 11. These addax photographed in northern Chad will prosper on arid lands where normal livestock cannot even survive; throughout its life the addax never drinks water but makes do with the fluids from the vegetation it eats. Photo: FAO.

Morocco, Algeria, Tunesia, most of Libya, Egypt and the northern Sudan are covered by the Sahara. The southern limit is around latitude 16 °N. This desert extends for more than 3,000 miles from east to west and up to 1,200 miles from north to south.

The desert is reasonably uniform in environmental conditions from east to west and from north to south. The topography, however, is rather varied and the desert is criss-crossed by mountain ranges, including the lofty Hoggar and Tibesti massifs. The highest peak of the Tibesti is over 3,600 m. The average elevation of the zone is less than 500 m. Great flat areas of stone and gravel sand and moving dunes occur throughout. Here and there clay sediments and salt deposits are present.

Climate

There are extreme variations of temperature both between summer and winter as well as day and night. In the summer the temperature difference between day and night is as much as 30 °C. The temperature may be as high as 45 °C. Rainfall is highly erratic.

About one-fourth of the total area has a precipitation of less than 20 mm per year and in most of the rest it ranges between 20 and 50 mm. Winter rains fall in the northern Sahara, summer rains in the south, and the central part has an irregular rainfall. In recent centuries no apparent decrease in precipitation has been noted. The prevailing wind is the northeast trade or 'Harmattan'.

Vegetation

Contrary to opinion that the Sahara is a barren wasteland, most of it has some plant cover. The number of species found, is, however, the lowest in the world, excepting the polar regions. Vegetation depends on the nature of the land surface and upon the presence of surface and subsoil water. Perennial plant life is confined mainly to favoured habitats, such as wadis, depressions and mountains. Over most of the Sahara bulbous plants and spiny bushes, sparcely scattered, only produce vegetative growth after infrequent rainstorms. On the scattered highlands there is sufficient pasture for camels, sheep, goats and antelopes. Large shrubs and trees appear where moisture is most abundant. Halophytes (salt loving plants) occur in saline areas.

Effects of human influence

Considering the relatively sparse human populations of this zone the impact of man and his livestock is striking. This is, of course, related to the limited capacity of the environment to support either man or his domestic stock. Human influence on the vegetation is apparent everywhere and is caused by the gathering of firewood for cooking and heating, lopping off tree branches for goat feed, or through the grazing and browsing of livestock. The grains, roots and bulbs of certain plants are also used as a source of food. Destruction of the flora has considerably accelerated in recent decades, and ultimately desert conditions will be created. Destruction of the natural vegetation leads to intensive wind erosion, which may accelerate the formation of dunes and consequently further destruction of vegetation.

Such forest stands as did once exist are now mere remnants occuring on a few isolated massifs or oases. Only two hundred Saharan cypresses (*Cupressus dupreziana)* survive in their original habitat, the Tassili Plateau of the central Sahara (STEWART, 1969). The mountain massifs of Ahaggar, Aïr and Tibesti are also plant refuges. The medemia palm (*Medemia argun*), once apparently common in Ancient Egypt, survives in two uninhabited oases (BOULOS, 1968).

NOMADISM

More than one million nomads live in the Sahara tending their herds which provide milk, butter, cheese, hides, wool and hair. Meat is unimportant, as nomads only slaughter stock for exceptional reasons, such as a religious festival. Camels are superior to other domestic stock in the desert in that they remain for a long time without water. This allows them to spread out widely and to use the scarce plant resources to a maximum. Nomads are faced with a varying supply of both pasture and water from year to year. The risk of drought and disease is high, constituting a certainty to each herdsman that there will be several very bad drought spells within his lifetime. In the Sahara livestock do not convert easily into money, which itself is not easily convertible into food. Economic surplus therefore can only be invested in more livestock which carries obvious hazards for the owner. Some nomad pastoralists, for instance in the Sudan, also partly live on the collection of wild plant products.

AGRICULTURE

Agriculture is restricted to the many scattered oases where date palms are mainly cultivated and where limited cereal and vegetable production is also possible. Other fruit trees in the northern oases include almond, apricot, fig, orange and olive. Some medicinal plants are collected in the desert for export, among which henbane (*Hyoscyanus muticus*) is the most commonly used. It has now almost been eliminated in the wild.

IRRIGATION

The discovery of large underground reservoirs of fresh water has opened up new possibilities for development. Where large enough underground water reservoirs exist to sustain irrigation, projects have been started and a very large project has been established in Libya. Another large irrigation project is at the Sudan Gezira on the Nile.

Reafforestation

Shrubs and trees play a very important role in the lives of the pastoral people, because in their absence stock raising would be virtually impossible. Reafforestation has taken place on extensive areas of sand dunes in Libya and in Egypt with *Acacia*, *Eucalyptus*, *Robinia*, *Tamarix*, poplar and pine with moderate to good success.

Wildlife

Considering the scarcity of the vegetation and the lack of water, the variety of species and numbers of wildlife present in this zone is indeed impressive, although due to man's activities the range and number of many species has much diminished. This is especially the case for several large carnivores like lion and cheetah, and for large herbivores like the addax, oryx, hartebeest, and ostrich. The development of oil fields and the presence of military transports has recently introduced new detrimental effects on wildlife. The wild herbivores could continue to be of great importance as a supply of protein which would otherwise not be available in the drier parts of the desert, as livestock, other than dromedaries, do not survive there.

Prospects

Effort on an international, regional scale should be made to stop further deterioration of the already sparse vegetation, particularly the shrub cover as part of the programme to reverse the process of desertification'

Many efforts are being made to solve the problem of spreading sand dunes. These include the application of oil sprays to stabilize the soil, followed by the planting of trees and shrubs. Further large-scale reafforestation projects should be undertaken wherever possible and these should include the establishment of shelterbelts in most cases. The large quantities of underground water which have been discovered in the western part of the Sahara will no doubt be tapped for irrigation purposes. The irrigation projects should be carefully planned so that no environmental damage follows development.

The Mediterranean zone

Geographic Distribution

The Mediterranean zone covers a relatively thin strip of North Africa, typified by a Mediterranean climate and vegetation, starting in Morocco and following the Mediterranean shore to Egypt. The strip is narrowest in Libya and Egypt where it is interrupted over large stretches by the Sahara. In the south west it is separated from the Saharian zone by the Atlas Mountains.

Climate and Physical Conditions

The climate is typically dry and hot, characterized by low annual rainfall which occurs mostly during the winter. Hot winds from the Sahara are frequent during the summer. To the south the Mediterranean climate merges with that of the desert climate of the Saharian zone. Within the sub-humid areas, which occur on the northern slopes of the mountain ranges, there are a multitude of variations in physical conditions which arise from differences in topography and from the influence of mountain and sea. These physical characteristics, together with soil and geological formations, have largely determined the composition and distribution of the indigenous vegetation.

Vegetation

There is a great diversity in vegetation types depending on ecological conditions. In the more arid areas the typical Mediterranean flora consists of evergreen trees or a predominance of perennial grasses and woody species. Forest types also vary greatly. In the more humid mountains evergreen oak and chestnut occur. 'Maquis' or 'Macchia' (Mediterranean shrub) is represented by numerous variants in climatically suitable high mountain country. Important stands of Aleppo pine (*Pinus halepensis*) occur there also. Another pine, *Pinus pinea*, is cultivated mainly for its seeds. Forests of *Juniperus* are important in Cyrenaica and Morocco. Also present are forests of *Callitris*. At the highest elevations cedars and firs (*Abies*) occur. Cork oak is grown for cork production mainly in Algeria.

Effects of Human Influence

Under Roman domination, agriculture prospered and became diversified and in the course of six centuries North Africa became one of the principal granaries of Rome. Grain culture, vineyards and olive-tree plantations extended over several million hectares of land and most of the best semi-arid land was intensively used by the Romans for several centuries to produce grain, grapes and olives. But as a result of mismanagement of the land (for example, through excessive continuous cultivation), food production decreased and by the time of the Mohammedan conquest in the seventh century, the land was probably in very poor condition, because of the nomadic habits of the Arabs. But the principal reason for the decreased productivity can be attributed to erosion (Le Houérou, 1970), a natural consequence of mismanagement.

As man learned to clear the forest, he pushed his fields even further into the forested area of hilly districts where the soil was richer in humus and rainfall more ample. Although the crops were naturally greater, soil exhaustion soon took its inevitable toll and heavy rainfall produced soil erosion gullies and eventually rendered cultivation impossible. Ground water has since seeped into the gullies and lowered the ground-water table. Much formerly forested land has been denuded. OEDEKOVEN (1970) mentions for example, that there are a number of indications that in ancient times parts of Egypt were covered with natural forest. The depletion of forest reserves gradually led to the disappearance of all sources of wood supplies in the Nile valley, the desert fringes and wadis, and also along the Mediterranean coast which once carried a light forest cover. Both in the hills and on the plains north of the Atlas mountains the original forest vegetation consisting largely of oak, chestnut and pine trees has been virtually eliminated.

There is no doubt that soil erosion, much accelerated by destructive wars, was one of the chief causes of the decline of agricultural production. Unfortunately, this process is still going on at an accelerating rate. It is terrifying to realize that man has been carrying out his destructive activities in this zone without forethought or let-up for something like 9,000 years.

AGRICULTURE

Agriculture which has had to contend with a low and uncertain rainfall and often with sloping terrain and poor soils, has remained the predominant activity in this zone. Generally, dryland farmed grain crops are interspersed locally with non-irrigated tree crops.

Agriculture in rain fed areas is largely confined to the production of cool season crops in the winter, wherever the rainfall normally is sufficient. Thus the range of crops which can be grown is severely restricted. One direct result of planting grain crops more extensively is the destruction of grazing land and the replacement of forage plants by annual weeds of low feed and erosion control value. Increasing population pressure has resulted in further degradation of the soil, and the effects of erosion, particularly on mountain slopes denuded of their forest cover, are increasingly apparent.

An uneasy partnership prevails between livestock raising and crop cultivation and true mixed farming such as is widely practiced in Europe is seldom found. Usually, hard wheat and barley sown in November-December is harvested in April-June, and the land remains fallow until the next crop is sown. In the poorer and drier areas the fallow may remain uncultivated for a number of years. During this time it is grazed. In addition to hard wheat, soft wheat and oats, pulses and some forage crops are grown. Very often these crops are mixed with scattered groves of olives, carobs (*Cerobania siliqua*) and almonds. Olive cultivation is a monoculture restricted mainly to the steppes of Tunisia but as the soil is kept bare, erosion occurs in the orchards. In the foothills and mountains viniculture is of increasing importance. The area sown to summer crops such as sorghum, cotton and maize, largely depends on an adequate rainfall in late winter to insure sufficient soil moisture for subsequent growth. Low yields are a major problem.

Land resources have so badly deteriorated that there is an urgent need for soil restoration through practices such as contour terracing. Farming should be done on terraces as much as possible in order to reduce soil losses and fodder tree plantings should be established for the production of animal feed. It is also possible to cultivate sown perennial pastures on terraces. Contour terracing, together with the planting of perennial grass and legume seedings on them has been remarkably well developed in certain parts of North Africa but it needs to be more widely applied.

Forests

Forest grazing of livestock has been practised for centuries in combination with man-made fires, resulting in serious forest degradation throughout the entire zone. Grazing naturally leads to changes in the forest composition. Various stages of degradation are recognizable, depending on the degree of exploitation and destruction. Through lack of natural regeneration, due to grazing, forest stands tend to become 'even-aged', namely consisting of over-mature trees.

Deforestation is proceeding rapidly, partly because increased agriculture in marginal lands is reducing the area of grazing available to stock and thus increasing grazing pressure on range and forest. Continuous exploitation of the forests, without sufficient conservation or reafforestation to insure the existance of adequate future supplies. has caused a rapid consumption of forest resources. The forest once seemed an unlimited source of various products, but wood was cut, the bark taken for cork or tannin, the fruit eaten, and the resin tapped. As populations increased, more and more of these products were extracted. The most valuable species were the first to be eliminated. As a result, forests have dwindled both in area and quality. Many of the forests which were overexploited turned into bushy stands, and eventually into open shrub and scant forest lands, where the soil has become so eroded that bare rock and subsoil are all that now exist.

The most commonly degraded forests are generally the open forests of Aleppo pine (*Pinus halepensis*) which occur in various stages of degradation. Phoenicean juniper (*Juniperus phoenicea*), rosemary (*Rosmarinus* spp.), alfa grass (*Stipa tenacisima*), cist (*Cistus libanitis*) and other forest relicts are often present (Le Houérou, 1970). The production of Arabic gum from *Acacia senegal* also contributes towards vegetation degradation. The steppes of arid North Africa, lying between the isohyetes 250 and 400 mm, originate from the degradation of a forest consisting of *Pinus halepensis*, *Juniperus phoenicea* or *Tetreclinis articulata*.

Rangelands

Ranges are rapidly being depleted due to overstocking, increase of temporary cultivation of cereals, and the uprooting of woody species for fuel. These phenomena are direct results of increases in the human population which has multiplied six-fold in the Mediterranean zone since the beginning of the century.

Photo 13. Typical Rif mountain landscape in Morocco: dry stream bed with cultivation on the slopes, eroded soil, and clearing of natural vegetation by burning, a practice which rapidly causes erosion in originally fertile areas. Photo: FAO.

Photo 12. Natural forest of green oaks *(Quercus ilex)* in the Middle Atlas of Morocco. Photo: FAO.

With the decline of natural grazing, and with folder cultivation not developing rapidly enough to replace it, increased livestock production has been possible only at the expense of a decline in the quality of natural grazing and of further inroads into the forest. LE HOUÉROU (1970) estimated that for the whole of North Africa about 100,000 ha of grassland are annually being rendered virtually useless by overgrazing. Reduction in stocking rates by as much as 90 per cent are needed over wide areas to bring the numbers of livestock down to the carrying capacity of the reduced area of land that should be used for pasture.

ANIMAL HUSBANDRY

Livestock constitutes the principal agricultural resource of the arid regions, but its husbandry is conducted in a backward manner. Some animals are kept alive too long and the excessive proportion of males serves no useful purpose from a production point of view. Livestock raising is extensive but little or no feed is stored; consequently, periodic disasters follow drought years. In the arid parts of the zone, more than 80 per cent of the land is available for grazing; this includes fallow fields and grain stubble. The rangelands are used continually without rotation and without the provision of feed reserves. There is practically no interchange between the raising of livestock in the steppes and fodder production in the irrigated areas (LE HOUÉROU, 1970) and the importance of growing fodder for livestock in irrigated areas has not been adequately recognised. Expansion of irrigated fodder crops should permit stock-piling of feed for the disaster years and supplementary feeding of lambs and other herded animals. It is estimated by LE HOUÉROU (*op. cit.*) that 30 per cent of the irrigated lands now used for the production of human food should be diverted to the raising of sufficient fodder for livestock. In view of prevailing food scarcities this will be difficult to accomplish. Mainly because of the extremely unfavourable feed conditions, the quality of livestock is very low and has not been improved. Indigenous breeds tend to be hardy rather than productive and their yields are low.

The planting of spineless cacti (*Opuntia ficus-indica*) has provided additional food for livestock, which can consume daily up to 10 per cent of their weight of this fodder, provided that they receive additional feed. *Opuntia* plantations produce from 10 to 50 metric tons of green feed per hectare per year, which is rather substantial considering the arid conditions. Saltbushes (*Atriplex* spp.) give excellent results, especially in non-irrigated areas. They are very resistant to drought, produce well, have a high feed value and a good palatability. A rotational exploitation of rangelands includes control of grazing pressure and of erosion, rotation and periodically deferred grazing, and availability of reserve feed (saltbush, cactus, hay and concentrates).

Erosion

Centuries of mismanagement of the soil have occurred in this zone. In the limestone massifs the steeper slopes have been completely denuded. Where the slope of the land brought under cultivation exceeds 3 to 5 per cent, continuous cultivation will lead to its abandonment in ten years. Removal of vegetation has exposed the soil to harsh physical conditions with subsequent rapid soil destruction. High temperatures on bare soil induce decay of humus and this reduces its cohesive capacity and ability to retain moisture. This in turn speeds up the movement of surface water and accentuates the erosion which follows, mainly from steep terrain. As a result of the more rapid water movement, floods have become torrential and of short duration causing heavy silt loads, and damage to the lower reaches of the waterways is intensified.

Wildlife

Wildlife has been depleted more in this zone than in the rest of Africa, beginning with the Greek and Roman occupations of North Africa. In Roman times large numbers of African animals, including lions and leopards, were taken to Rome for use in public sports. It is also certain that in the recent past more typical Ethiopian mammals were present on the southern shore of the Mediterranean than is now the case. Most of these mammals seem to have disappeared from Egypt during the late Roman era, although the hippo lingered on in the Nile Delta until 300 years ago (Moreau, 1966). Buffalo, elephant, giraffe and hippopotamus disappeared from the Maghreb countries at dates varying from 2,000 to 300 B.C. (Moreau, *op. cit.*) The Barbary stag, a close relative of the red deer and once widely distributed in northern Africa, is now confined to a small area in eastern Algeria and western Tunisia.

Not only have many species of mammals become much reduced in numbers, this is also true for birds. For instance, a once wide-ranging bird, the waldrapp or hermit ibis (*Geronticus eremita*) is now restricted to small breeding colonies in Morocco and Algeria.

Prospects

Owing to population pressure, it is clear that more and more land will be cleared for agriculture. It is necessary to ensure as much as possible that the right land is chosen for this and not the unsuitable land that has already been cleared. It has unfortunately become increasingly more obvious that traditional agriculture and pastoralism cannot assure the population a sustained food supply. Agriculture in the arid parts of the zone should include the production of irrigated crops such as vegetables, fruits and flowers, and not solely cereals. More legumes should be produced on irrigated lands. With the application of improved technology there might be a possibility of increasing output to levels which would render unnecessary some of the present marginal cultivation. The purpose should not only be to arrest the process of soil destruction, but to achieve a long term increase in soil fertility by using land and water resources to the greatest possible advantage.

Although it would be difficult to accomplish under the present circumstances, as explained above, forests should be maintained where they exist; their density increased and their composition improved through silvicultural practices. Stands should be re-established at higher elevations by the use of ecologically adapted species. Shelterbelts should be established where needed.

Future management could well comprise of two main lines of action. Firstly, the protection of timber stands against grazing, and more particularly elimination of goat grazing and browsing from all forests except evergreen sclerophyll forests which can better withstand this pressure; the elimination of sheep grazing from fir forests, except in the open and woodland pastures included in the major forest area; and the elimination of any grazing from important watersheds. Secondly, management of inforest grazing elsewhere would be permitted in combination with forest-rangeland management and improvement.

These two objectives cannot possibly be attained unless they are supplemented with fodder production in the adjoining agricultural and rangelands. In addition to the fodder trees, numerous shrub-like forage legume species may well serve as another possibility towards creating new sources of fodder. Deciduous forests and shrubs still play an important role in supplementary feeding. The culture of fruit producing trees such as figs and olives is an excellent example of useful, large scale afforestation. There is also good potential for pistachio cultivation in the arid high plains of Algeria and Morocco, which should serve the dual purpose of increasing both the income of the farmers and protection of the soil. Afforestation with exotic species of *Eucalyptus* and *Acacia*, which has already taken place on an experimental basis, should be initiated on a large scale.

The outlook for food does not look good. Even if draconian measures are taken to improve the situation, these efforts will be largely nullified by the rapidly increasing population. Famines can therefore be expected in this zone during the next 10 to 20 years.

The Eastern zone

Geographic Distribution

The Eastern zone covers most of what is commonly known as East Africa: Ethiopia, Kenya, Uganda, Somalia and most of Tanzania. The greater proportion of this zone consists of highlands which assume their most rugged form in Ethiopia.

Climate

The main features influencing the climate are that the zone is situated astride the equator and that much highland exists.

Within this zone, climatological and weather conditions largely determine the existence of a series of ecological systems which vary from desert, with extremely low annual rainfall, to evergreen equatorial with extremely high (2,000 mm) annual rainfall. The intensity and distribution pattern of rainfall is a dominant factor in determining vegetative growth. As the average annual rainfall for this zone is less than 875 mm the chance of serious drought is high. The duration of the dry season depends on the latitude. The north-south climatic belts in the zone vary with the position of the two wind systems, the south east and north east trades. Most of rain falls in the transition period between those two systems.

Vegetation

The vegetation pattern closely reflects the distribution of rainfall and is also influenced by the broken topography of the zone. This is the most diversified zone from an ecological point-of-view. In the uplands along the water courses and around Lake Victoria, there is tall and dense forest, often with mahogany (*Enantophragma* and *Khaya*) prominent, or with *Celtis* at higher elevations. The mountains are covered with evergreen forests, including the native 'cedars' (*Podocarpus* and *Juniperus procera*). Above these are dense bamboo brakes merging at higher elevations into *Hypericum* scrub and the unique African alphine flora of giant *Senecios*, *Lobelias* and heaths. According to the vegetation map of Africa about half of this zone is covered by wooded steppe, while the dryer portions are occupied by subdesert steppe – tropical type. The most commonly observed genera consist of *Brachystegia*, *Commiphora* or *Acacia*, the trees becoming smaller and more widely spaced as the rainfall diminishes. In the driest areas only sparse, semi-desert scrub now remains, for frequent fires have prevented the regeneration of trees.

Effects of human influence

The destruction of the vegetative cover by man, and the consequent degradation of the soil, began probably in the prehistoric ages, the first stages being the destruction of the wooded vegetation. Cultivation of clearings resulted in swift deterioration of the soil. Where the clearings were destined for grazing, they were invaded by grasses and the continuation of grasslands was guaranteed by periodic fires. The successive stages of degradation of the vegetative cover and of soil deterioration are particularly obvious in the arid and semi-arid regions. Deforestation, fire and overgrazing pressures continue to be applied with ruthless and unconscious perseverance.

Forests

Forests have been cut almost everywhere and remnants of lowland forests survive only when they are protected as reserves. Here and there on the mountains evergreen forests still occur. Even these reserves are constantly in danger due to fire encroachment or illicit felling and removal of saplings for house poles and other purposes.

The amount of deforestation which has taken place during the last half century is most serious. In Somalia, for example, deforestation is particularly pronounced and regeneration of the relic Juniper (*Juniperus procera*) has been either unsatisfactory or absent for some years now. The mangrove forests of southern Somalia have disappeared almost completely as a result of over-exploitation, and migratory sands are menacingly approaching the important agricultural and population centres near the coast.

In Ethiopia, most of the *Juniperus procera* forest type, which may share dominance with *Olea chrysophylla*, has been replaced by degraded montane shrub. All the forests of Ethiopia are threatened because of rapidly increasing commercial utilization of their timber. Little, if any, rainforest remains in virgin condition in the country.

In Kenya and Tanzania, with encroachment into many inadequately protected mountain reserves occurring, the forest cover of many mountains is rapidly receding. In Tanzania the *Brachystegia* forest is also much modified by man, and probably the only climax woodland of this type remaining is on steep and rocky escarpments, or in tracts of country remote from domestic water supplies.

The trees in most cultivated landscapes are introductions, particularly *Eucalyptus* spp. and wattle (*Acacia* spp.), both from Australia and both growing with phenomenal speed. 'Cedars', especially *Juniperus macrocarpa*, are also often planted with success.

Agriculture

The interaction of climate and soils largely determines the pattern of land use. Elevation also has an important effect on agriculture. The range of crops, grass and legume species which are available for agriculture in this zone is great. Most East African countries produce similar staple food crops: grains such as maize, sorghums, millet, wheat or rice, supplemented in varying degree by pulses, root crops, vegetables and fruits. The most fertile highland areas allow farming on standards close to those of Western Europe.

The following principal geographical regions may be distinguished from the point of view of suitability for agriculture:

Photo 14. Wildlife on open grasslands. Hartebeest in Masailand. On the left is a warthog. Kenya Ministry of Information photo, issued by FAO.

1. The Eastern Equatorial Desert and Sub Desert is composed of a large part of Somalia, the hinterland of Ethiopia, and the northern region of Kenya. This region is arid and the quality of the soil is poor. It is suitable mainly for livestock grazing.
2. The Eastern Equatorial Savanna extends from Ethiopia through Kenya to the heart of Tanzania. Rainfall is moderate but diminishes as one moves northwards. This area is used mainly for grazing, but is suitable for the growing of millet and sorghum.
3. The Eastern Coastal Region consists of the unbroken chain of a narrow coastal strip stretching southwards from the lower part of Somalia to Mozambique. With the exception of the Somalia segment, abundant rainfall prevails throughout. The major crops found in this region are cassava, rice, cashews, coconuts, bananas and sisal.
4. The Lake Victoria Region constitutes the most fertile and most heavily populated section of the zone. The soil is basically of the alluvial type which is highly productive and the annual rainfall is abundant. The staple crops are maize, cassava, plantains, pulses and vegetables, and the major cash crops include robusta coffee, cotton and tea. Livestock does not thrive well in the region because of the presence of tsetse fly.
5. The Eastern Plateau comprises the largest part of Tanzania. The soil is fertile alluvial and cotton, tobacco, maize, millet and sorghum grow well.
6. The East African Highland is widely dispersed in Ethiopia, Kenya, Ruanda and Burundi. The climate is cool, the soil generally fertile, and the annual precipitation high. The basic food crops include maize, sorghum and pulses. Coffee, tea and pyrethrum are raised primarily for export.

The most important sphere of agricultural activity in East Africa is crop production. Coffee is the principal agricultural export but tea also is an important product of which Kenya is the main exporter. Kenya and Tanzania are the principal sisal producers, although in recent years many plantations have been discontinued due to a considerable drop in the price of sisal. On the other hand, cotton production has increased considerably. Linseed is grown in Ethiopia and Kenya, sunflower seed in Ethiopia, Kenya and Tanzania, while sesame seeds and castor seeds are widely produced throughout. Tobacco also is widely grown. Groundnut and coconut production is a fairly important industry. Sweet potatoes and yams provide a major contribution to the human diet in East Africa and cassava is here regarded more as an emergency food than as a staple.

The original native systems of agriculture were also based on shifting cultivation within the forest but the pressure of increasing populations has meant that a satisfactory rotation is now no longer possible. At present the optimum procedure is to allow elephant grass (*Pennisetum*) to recolonize this fallow land. Although this grass fallow does improve subsequent crop yields, it is unlikely that it adequately maintains the fertility status of the soil. In many places the population density has become so high that even grass fallows are impossible. Consequently, crops show increasing signs of soil deterioration and nitrogen starvation.

There is abundant evidence from heavily settled areas that the soil has become increasingly more exhausted. The most urgent land use problems to be solved in Tanzania's mountains are in the densely populated agricultural areas outside the forest reserves. Here, the sharp increase in population during the last few decades has caused an obviously higher pressure on the land resources. Steeper slopes are now cultivated, the period of fallow becomes shorter and eventually non-existent, and the existing remnants of forest and woods are cleared. In Ethiopia even mountains as high as 4,000 m are almost completely cultivated and are entirely devoid of trees. On many soils excessive land use has caused accelerated erosion of different types: soils and montane agricultural potential are destroyed, and the quality and quantity of the lowland water flow in rivers emerging from these mountains is reduced. The compaction of the surface soil due not only to unstable vegetation, but also to damaged soil structure caused by livestock trampling, results in greatly increased surface run-off.

Rangelands

The natural grass cover varies considerably from the highly productive *Themeda* of Uganda, Ruanda and Burundi through the productive *Pennisetum* types of Kenya-Tanzania to the less productive types of Ethiopia-Somalia.

In most parts of East Africa the range has been seriously depleted as a result of continuous grazing pressure by livestock. Overgrazing by domestic animals has become a rapidly increasing problem due to better disease and parasite control and consequently, steadily increasing herds. In Somalia and Ethiopia particularly, the range has suffered greatly from overstocking. Many places described by early travellers as pleasant park like country inhabited by herds of game, are now grim wasteland cover. In Ethiopia even high mountain ranges are locally overgrazed.

Semi-arid lands have suffered greatly from attempts at ranching: the destruction of the original woodland cover, long continued overgrazing and frequent burning have practically removed the original deep rooted vegetation. As a result, the vegetation has diminished in productivity and has become more open, bare soil being exposed between the plants. The original range is capable of supporting cattle but as it becomes depleted it supports mainly sheep and finally only goats and camels. In the drier parts of the zone, such as eastern Karamoja, Uganda, as well as parts of Kenya and Tanzania, only semi-desert remains, although an annual rainfall of 750 mm or more should normally support a fairly dense woodland.

The deterioration is evidenced by a thinning of the vegetation and an increase in the proportion of unpalatable or harmful species. Areas of previously good grazing have become stretches of loose pebbles and bare ground, and ever increasing amounts of water are lost in run-off. Near permanent water points the terrain is often completely denuded.

Photo 15. At least four-fifth of Kenya is rangeland. The Government has an ambitious integrated plan to develop a highly organized livestock industry with export potential. An important objective is to settle the nomadic Masai tribe on so-called Group Ranches. An unusual sight in Masailand are sheep. In the background is a water-tank, one of many that will one day dot the often – parched rangeland. Kenya Ministry of Information photo, issued by FAO.

Overgrazing is perhaps most serious in Somalia where trees and bushes have been used for fodder, fuel and building material, and overgrazing has destroyed the grass. Remnant water supplies in the forest areas have dried up. As a result of extensive veterinary improvements the herds of domestic animals have grown even larger despite warnings about the consequences. Provision of permanent water supplies throughout the country has meant that areas formerly having a seasonal respite from grazing are now more or less continuously grazed.

There are indications that the Somali himself is becoming conscious of these problems, but unless immediate action is taken, the whole of the grass steppe region will soon be a waste land (VERDCOURT, 1968). The forest remains open to grazing and much damage has resulted from erosion and lack of regeneration. Even succulents are suffering either due to direct grazing or lack of sufficient root support due to wind erosion; those of an unpalatable nature are greatly increasing at the expense of the palatable species.

Animal husbandry

Within this zone the human population is divided into two groups: a minority who occupy huge areas, maintain large herds and should have sufficient or abundant animal protein in their diet, and a majority who are agriculturists, possess few livestock, and are chronically deficient in animal protein.

Much of the drier country is occupied by cattle raising populations like the Masai and the Karamoja tribes who live almost entirely on diets of blood and milk derived from their herds. Where the rainfall and vegetation are strongly seasonal, the tribes are nomadic.

Although Somalia is inhabited mainly by pastoral people, its livestock resources are inadequate to support its people. It is not surprising, therefore, that much of the country is badly overgrazed.

Several countries in the zone still export beef but this is not likely to continue considering the needs of rapidly increasing domestic populations. Attempts to increase beef production have been hampered by inadequate knowledge of the livestock and grazing resources and by ownership customs by which cattle are considered as a measure of wealth and not a means of production. However, as a result of extension and demonstration work, the attitudes of nomadic people are slowly changing in favour of a monetary economy and consequent stock improvement.

Wildlife

One of the most valuable natural resources of this zone is its varied wildlife populations, which are mostly herbivores. Although diminishing, the populations are locally still extremely large and have drawn world-wide attention because of the great variety in number of species and animals. Recent (1967) estimates by various biologists indicate that the Serengeti-Mara region of Tanzania-Kenya (24,000 square km) alone contained more than half a million Thomson's gazelle (*Gazella thomsoni*), about a third of a million wildebeest (*Connochaetes taurinus*), 175,000 zebra, and several thousand eland (*Taurotragus oryx*), topis (*Damaliscus korrigum*), hartebeest, buffalo and elephant. While this represents one of the most outstanding concentrations of plains wildlife in Africa, other areas also contain relatively abundant numbers of wild mammals. It has been calculated by Petrides & Swank (1965) for example that the total weight of biomass production of wildlife in the Ruwenzori National Park in Uganda amounts to 27,8–31,5 T/km^2, a figure which compares favourably with production of beef under intensive management.

Wildlife is most endangered in this zone in Somalia. Predators like lion, leopard and cheetah are disappearing due to poaching and poisoning. The dibatag (Clarke's gazelle) is disappearing due to the removal of *Commiphora* trees, its main fodder.

Forestry

Less than ten per cent of the forest land of the Eastern zone is covered by closed high forest. This forest, together with man-made forests represents an important resource and a potential base for forest industries development.

Shifting cultivation has taken a heavy toll of the high forests in the past and although these forests are now to a large extent protected through reservation and other means, shifting cultivation continues in large areas of Ethiopia and Kenya.

Since a large part of the zone has areas of high elevation, low population density, and ecological conditions that are unusually favourable for the establishment of exotic forest plantations, these are now widespread throughout the region. Coniferous plantations already constitute an important source of sawlogs, fuelwood and small roundwood for domestic and agricultural use. *Varsis* and *Cupressus* spp. are mainly used for these purposes but optimistic yield forecasts have been threatened recently by setbacks in the form of insect, wild mammal and fungus attacks on a large scale. The principal reason behind these attacks is the ecological change caused by the replacement of a strongly heterogeneous, mostly non-coniferous natural forest ecosystem by introduced softwood monocultures. Less visible, but in the long run probably more serious, are the negative changes which can be expected in soil fertility as a result of continuous cultivation of these conifers.

Prospects

In agriculture the main problem will continue to be the checking of erosion and the maintenance of soil fertility. Mixed farming should be more widely practiced and better use should be made of the available manure. Soyabeans should be concentrated on because of their high protein content. Many improvements are required in range and pasture management and the start with improved pastures on fenced rangeland should be more widely applied. Productivity per animal unit could be considerably improved if contagious and other diseases were better controlled as they are in Kenya.

Tourism has become the largest source of foreign income in Kenya and the spectacular game herds are the backbone of the tourist industry in East Africa. In spite of its potential, the future of wildlife is uncertain because of human population pressures, the rapid rate of adjudication of tribal and trust lands to private ownership and the traditional non-commercial role relegated to wildlife by the public. Plans for the management of wildlife in national parks and other wildlife areas should be carefully coordinated with national land use plans.

The Zambezian zone

Geographic distribution

This zone consists largely of high plateau country situated south of the Guinean and Eastern zones and north of the Transvalian zone, covering Zambia, Malawi, most of Rhodesia, Mozambique and Angola, southern Zaïre and southern Tanzania. The plateau drops from an average height of 1700 m in the east to some 800 m in the west. In the east several mountains rise to heights of 3000 m. Here also there are two arms of the Rift valley.

Climate

The great diversity of height is reflected in wide ranges of rainfall and temperature. The largely subtropical climate approaches tropical conditions in the north. In the lowlands below 1000 m high temperatures and high humidity accompany the summer rainfall. In the highlands where the temperature is lowered by altitude frosts occur occasionally in the dry season. Rain falls mainly during the period December-May and decreases gradually from the east coast westward, with the exception of the highlands.

Vegetation

Due to wide ranging altitudes and related climatic conditions this zone has great ecological variety. The natural vegetation is mainly woodland – savanna, varying in luxuriance with available water supply from a rich flora in the highlands in the east to a rather impoverished flora in the west. Forests with a composition approaching tropical vegetation follow the courses of the principal rivers. Two main forest formations, the Miombo and the Mopane (*Colophospernum mopane*) woodlands, occur in the zone, the latter at lower elevations. By far the greatest portion is covered by Miombo which consists of relatively open stands dominated by *Brachystegia* and *Julbernardia* species. *Borassus*, *Hyphaene*, *Acacia* species and the Baobab also exist. This forest formation covers the southwestern half of Tanzania, the eastern halves of Rhodesia and Angola and about 80 per cent of Zambia. Mopane woodlands generally occur in the hotter and drier parts of the zone, often in river valleys. The dominant species is the mopane and there are many associated species, such as *Combretum* and *Terminalia*. Most river valleys are flooded for several months during the wet season and support productive grassland communities dominated by *Echinochloa* and *Vossia* species. The grass cover of the plateau is varied and where the tree story is not too dense it can be productive.

The first cattlemen in this zone not only took advantage of the grasslands and wooded-savanna vegetation, produced and maintained by fire during the Stone Age, but their shifting, slash-and-burn methods of cultivation further accelerated the trend towards more open grassland and wooded-savanna. Because of the shifting nature of their cultivation, a comparatively small population living in closeknit communities and occupying relatively small areas for short periods was able to produce profound changes over very extensive areas (WEST, 1972).

The swamp forests, which must have covered many of the presently open swampy areas, were almost entirely destroyed by the activities of prehistoric man, who, because of the wetness of the soils, was able to grow crops in these areas during the dry season.

Even before European settlement started, spectacular changes in the forest vegetation had already come about. Tropical and montane forests of the high rainfall areas in the east had been cleared and over enormous areas hillsides formerly covered by evergreen forest had been terraced and cultivated. Only relict patches of these forests now remain. There is also much evidence to show that the present widespread occurrence of *Brachystegia* woodlands has resulted from the effects of man's activities (WEST, 1972).

Much damage was done to the land during the early phase of colonization when the settlers, assuming that there was an unlimited supply of good land, made an attempt to consume it. It is estimated that in some of the most fertile regions several feet of topsoil have actually been removed (HAILEY, 1957).

In the more arid parts of this zone, such as Matabeleland in Rhodesia, similar observations of land degradation as in other arid parts of Africa can be made. The range around waterholes is gradually drying up for the first time in living memory (PRESCOTT, 1972). Overgrazing and excessive burning have caused severe sheet and gully erosion, particularly in the western part of the zone.

AGRICULTURE

On the more fertile soils agricultural production can be very good and high yields of grain crops are obtained under good management. Farming is centred around the cultivation of maize, millets and groundnuts. Sorghum and cassava are grown as main food crops. A wide range of vegetables and

---

Photo 16. Elephants browsing. Unlike cattle which concentrate on eating mostly grass, often overgrazing the land, each wild species has its own habitat and feeds on different parts of the vegetation. Elephants may push over trees, making the higher branches available to smaller animals. Photo: FAO.

fruits are produced in the zone, and when irrigation is used, especially good yields are obtained. Tobacco, cotton, tea, and citrus fruits are grown mainly under plantation conditions. Cotton is of little importance so far, but its cultivation could be greatly extended. Pilot schemes with the growing of rice have been successful.

Animal husbandry

Livestock production, particularly of cattle, but also of sheep and goats, is important in many areas. Many well managed ranches occur in Rhodesia. Extensive areas are still closed to cattle production, because of tsetse fly. Pigs and poultry are produced locally with considerable success, indicating that these industries could be expanded.

Livestock production can be stepped up considerably by better pasture management and the establishment of mixed farming. In Rhodesia, mixed farming based on sown pastures and locally grown concentrates such as maize and cottonseed meal is becoming increasingly important.

Range management

Because of increased grazing pressure grass that was formerly burned is now eaten, fire can no longer exercise its role in producing and maintaining grassland, and so the bush thickens while the grass decreases (West, 1972). Bush encroachment resulting from overgrazing and poor management is most marked in the regions of medium to high rainfall. In the drier regions, where it proceeds at a slower pace, grazing pressure has led to the denudation of millions of hectares of grazing land and most of the perennial grasses have been virtually eliminated. Conditions in some areas are steadily getting worse, and this decline is reflected in the changing balance of the livestock population because the degradation is imposing severe limitations on the growth of the cattle population but not on the goat and sheep population (West, 1972).

Forestry

The Miombo woodlands are still not intensively utilized. They are generally managed by clear-felling, followed by natural regeneration with supplementary seeding and planting. They furnish fencing posts, firewood, mining timbers, and flooring and around cities much forest land is exploited for the production of charcoal. Honey and beeswax are important by-products of Miombo woodlands, and are mainly used for local consumption.

In north west Rhodesia and south west Zambia there are broad ridges of

---

Photo 17. Acacia albida stand in the Luangwa, Zambia. Note the overgrazing of the herbacious layer and the damage done to the bark of some trees by elephants. Photo: A. de Vos.

dry deciduous woodland characterized by the commercially important Rhodesian teak (*Baikiaea plurijuga*), Rhodesian mahogany (*Guibourtia coleosperma*) and *Pterocarpus angolensis*. The latter species has a wider distribution. The stands of Rhodesian teak have been almost depleted. The Mlanjé cypress (*Widdringtonia whytei*), found in the highlands of Malawi and Rhodesia is a most useful tree for lumber production. Large conifer plantations have been established in the montane areas. There are extensive eucalyptus plantations in the eastern and central parts of the zone.

Wildlife

The larger game species, such as elephant, hippopotamus and buffalo are now largely restricted to national parks and game reserves as a result of increasing population pressure and opening up of the land for agriculture and animal husbandry. The Government of Zambia has now embarked on experimental cropping schemes of elephant, hippo and Kafue lechwe (*Kobus leche*), although the latter may be threatened by the inundation of the Kafue Flats. The black lechwe, a related race, has a very limited distribution in Zambia and Zaïre. The giant sable (*Hippotragus variani*) has a limited distribution in Angola. Many national parks and game reserves exist in this zone and afford reasonable protection to most wildlife species.

Prospects

The clearing of land of tsetse fly infestation continues and this will release more land for agriculture and livestock production. Irrigation systems will be extended wherever possible. The introduction of better seed mixtures, including legumes, should substantially increase pasture production. The still largely undeveloped irrigation potential of the zone could provide growing-season grazing at high stocking rates on fertilized, irrigated pastures. Forest plantations should be extended and more use made of introduced species, such as *Eucalyptus* and *Pinus*. Increases in crop production can be achieved by the introduction of higher yielding varieties. Efforts should be made to grow rice in suitable areas.

The Transvalian zone

Geographic distribution

This zone is the most heterogenous of the ecological zones of Africa and perhaps requires revision of its boundaries. It stretches across a relatively

---

Photo 18. Baobab tree damaged by elephants in the Luangwa Valley, Zambia. Elephants sometimes remove so much of the enormous stems of these trees that they die. Also note the browsing by elephants on the smaller trees and shrubs. Photo: A. de Vos.

narrow strip along the west coast of Angola, then spreads easterly and covers its southernmost part and the northernmost part of South West Africa. Subsequently, it covers northern Botswana, southern Rhodesia, northern Transvaal and southwestern Mozambique, and then stretches southward, following the coastal plain of Natal (fig. 4).

Climate

The climate of this zone is also rather diversified. It can be subdivided into two main climatic sub zones, the east coast where moist, subtropical conditions prevail largely as a result of the warm Mozambique current; and the interior plateau which is colder and drier. Apart from the moist highland areas, annual precipitation is less than 25 cm. Frosts and cold winters are rare. The dry South East passat winds affect the climate of the eastern part of the zone, while the cold Benguela current affects the western part.

Vegetation

Most of the coastal belt of Natal was at one time covered by subtropical forest vegetation which has been largely cleared to make way for agriculture. However, *Podocarpus* and broad-leaved trees still cover the seaward slopes and ravines of the mountain ranges in western Natal. Northern Transvaal and the southern part of Southern Rhodesia are covered mainly by savanna woodlands with both scattered and dense trees, including mopane, baobab, *Commiphora* spp. and *Acacia* spp. Many productive grasses, including, for example, *Digitaria* spp. are to be found here. The broad semi-arid region across southern Angola and northern South West Africa is dominated in the west by a large tract of mopane and in the east by mixed *Acacia* savanna.

Effects of human influence

Although the eastern part of this zone has been affected by relatively intensive human land use pressures only during the last two centuries since the invasion of the Bantus and the Boers, the land has nevertheless been seriously downgraded and increasing desiccation is taking place. This is well described by PROTHERO (1972) as follows: 'The value of the veld is being reduced through overgrazing, which is upsetting the normal balance of vegetation at the expense of the more nutritious grasses. Around waterholes in the drier part of the country, land is being denuded of vegetation, and erosion is occurring. Some waterholes are becoming dry for the first time in living memory'.

Photo 19. Botswana is richly endowed with wildlife, but this resource has been threatened by increasing competition from livestock, particularly cattle. Open-billed storks in a game reserve in northern Botswana. Photo: FAO.

The importance of the following statement in the Desert Encroachment Commission Report is evident: 'Soil moisture, normally held in the upper layers, has, by a network of artificially created runways, been depressed to lower levels or completely drained away. Hundreds of dried out vleis*, through which deep erosion channels have been formed, are there to testify to the way in which soil water has been drained away. This process is moreover not confined only to the immediate vicinity of the vleis; for long distances away from watercourses the water has been drained to lower levels with very grave consequences to the vegetation'. This widespread deterioration of the vegetation is particularly noticeable in the more arid regions, where overgrazing, trampling, and veld fires have caused serious deterioration of the land.

Many areas which are now under thick *Acacia* scrub were once open grasslands. This is a consequence of virtual elimination of grass competition by overgrazing. In the northern Transvaal thousands of hectares of formerly productive land have lost practically all of their original topsoil, and only subsoil remains, the latter assuming the hardness of rock during the dry season.

Much land that had never been farmed before has during recent decades come into production. The dangerous effects of this development were first noticed when serious shrinkages in water supplies occurred due to the formation of gullies and the consequent rapid run-off of rain water. Rivers became more seasonal in flow and their silt loads increased. In Transvaal, practically all available water has been overtaxed. Thousands of drilled wells have dried up in the early parts of this century.

Agriculture

On the plateau there are many difficulties with agricultural production, including irregularity of rainfall, periodic droughts and soil erosion due to the torrential and seasonal character of the rain. Corn is the main staple food of the Bantu population, but these people also produce large quantities of sorghum. Winter wheat is the main crop on the European farms. Other field crops include corn, oats, barley, rye, potatoes, groundnuts and sunflower seeds. Tobacco is grown fairly widely.

Tropical fruits, such as papayas, bananas, pineapples and mangoes are grown on the coastal belt of Natal and mainly under irrigation in the lowveld of Transvaal and Rhodesia. Citrus orchards are found in widely scattered areas and are the basis of a flourishing export industry. Sugar cane is grown mainly in the coastal belt of Natal.

Practices in tillage farming have been poor for many years, but important improvements have occurred in recent decades, including a considerable increase in mechanization, such as the use of tractors.

* marshes

Forestry

Fires have always brought a great deal of destruction to the forest. Not only have some fires completely wiped out large tracts of forests which have never become re-established, but efforts have to be made to control burning of rangeland on the margins of forests. The subtropical forests of the coastal area yield several varieties of fine, hard timber. Plantations of exotic species, such as pines, the Australian black wattle and eucalyptus have been established to replace the low-production native woodland. Large-scale plantations of Carribbean pines have been established on the sandy Zululand coast.

Livestock production

Cattle, sheep and goat raising has been the principal interest of the people on the plateau for a long time. The coastal strip and northern Transvaal are among the finest cattle regions of South Africa. The beef industry is based principally upon Afrikander cattle, an indigenous breed of livestock well adjusted to local conditions and also a good draft animal. The main pastoral products are wool, mohair, cattle hides, sheep, lamb, goat and kid skins. Wool is the most important of these products and exported on a fairly large scale.

Range management

The average low carrying capacity of the land explains the great importance of extensive grazing and the predominance of pastoralism on the agricultural scene. There are many constraints to livestock grazing, including the low carrying capacity of the drier regions, losses sustained in drought years, the difficulty of improving the veld grasses, the prevalence of diseases in the hotter areas, and the scarcity of water supplies.

Wildlife

When the Europeans first came into the interior, enormous herds of game used to move across the country, but these were drastically reduced by overshooting. As a result of subsequent protection, wildlife can again be considered an important resource in this zone from a touristic and protein production point of view. Large national parks and game reserves have been set aside for the protection of wildlife and among these Krueger National Park in Transvaal is world famous. Outside these sanctuaries wildlife was largely on the wane until some ranchers started to recognize the value of producing selected herbivorous species for their meat and by-products on a commercial basis. An increasing number of ranchers in Transvaal and Rhodesia now maintain several species of wildlife, such as blesbok and impala, on their ranches.

Several species of wildlife, once endangered with extinction, have increased

greatly as a result of careful protection. The outstanding example is the square-lipped rhinoceros which increased from a very low population to more than 2,000 in 1971. Another example is the white-tailed gnu (*Connochaetes gnou*) which was saved from extinction by some farmers in the Orange Free State and now also exceeds 2,000 in the wild.

Prospects

Although an inadequate number of suitable sites are available for further storage of water, the amount of irrigated land certainly will have to increase if increasing needs for water to feed the people and their stock are to be met. In a thirty year plan the total mobilization of the Orange River has been outlined to obtain additional water for agriculture, industry and power. Moves are now under way to exploit the water resources of the Okavanga Swamp in north eastern Botswana. Hopefully the impact on the ecology of the swamp will be carefully studied before any development takes place.

A great deal more emphasis should be given to range improvement, to include a solution to the problem of providing for winter grazing, since few grasses retain their feed value in winter and fewer still will thrive when subjected to winter grazing. That the carrying capacity can be increased substantially is indicated by the work of West (1972), who calculated that over most of the southern lowveld of Rhodesia the existing cattle population could live on fertilized, irrigated pastures at a stocking rate of five head to the acre during the six months (November-April) of the rainy season. A number of range improvement practices will be necessary, including the protection of the catchment areas, the abolition of night kraaling, the control of veld burning, and grazing control to ensure the periodic resting of pastures.

The Basutolian zone

Geographic distribution

This zone encompasses the so-called 'High Veld', occupied by the southern part of the Transvaal, the Orange Free State, Transkei and Lesotho (fig. 4). It is formed mainly of a high plateau (1350–2000 m) which is rockier and hillier in its eastern part.

Climate

The climate is temperate continental and is characterized by dry, cold winters with severe frosts and hot summers. Below 2000 m the rainfall ranges

---

Photo 20. In 1897 there were only 20 white rhino left in the whole of Southern Africa. Now there are 2000 or more, many of them re-introduced into formerly depleted range. A group of white rhinoceroses. Photo: FAO.

Photo 21. Fine angora goats in Lesotho, a small nation surrounded by the Republic of South Africa. Although the range is overgrazed, good quality sheep and goats are being produced here.

between 62 and 75 cm. The variability of precipitation is a serious disadvantage to agriculture.

Vegetation

Most of this zone is covered by temperate and subtropical grasslands, the composition of which is influenced by soil conditions, grazing pressure and fire history. There are occasional shrubs and trees to be seen. Infrequent, small shrub communities occur which form a closed canopy under which very few herbacious plants can flourish. Most of the rocky hillsides carry Protea open woodland, in which *Protea caffra* is dominant (Davidson, 1964).

There is evidence that a great deal of forested land was changed by the Bantu into grassland along the eastern portions of Natal, East Griqualand and the Transkei.

Records of travellers of about a century ago included descriptions of well watered grassland in what are now regarded as the dry sections of the central Orange Free State and Western Transvaal (SCOTT, 1951). They also reported grasslands in the western part of the zone where karroo vegetation prevails now. Until recently the Bantu practiced shifting cultivation mainly characterized by bad land usage and inefficient agricultural methods: steep slopes were plowed and fertilizers were not applied. Yields were consequently comparatively poor.

As the population increased there was no more space for shifting cultivators to occupy and nothing was returned to the soil and no protection was maintained against erosion. The soil, instead of being properly farmed was mined, thus deterioration normally followed. Eventually, much land had to be abandoned to the ravages of soil erosion.

From the standpoint of the carrying capacity of the land the keynote is overpopulations of people and stock, particularly in the Bantustans. Widespread deterioration of the vegetation has occurred as a result of many decades of continuous grazing by livestock, followed by widespread loss of soil by erosion. After the removal of the more palatable and soil-holding grasses, there has been a marked spread to the east of the more arid Karroo-type vegetation (see p. 123) at the expense of grassland with a consequent reduction in carrying capacity of the land.

There is little doubt that the land continues to deteriorate. This is particularly true for the Bantustans where holdings are too small to permit modern farming practices or to support the farmer and his family. In Lesotho even the highest peaks have been grazed over and there is not a hectare of ground within the mountain area that is entirely free of grazing the year round.

There is a fairly high rainfall in this area and peat-bogs, which occur here act as filters, and provide a steady flow of clear, silt-free water to the streams and rivers. But with increasing grazing pressure, there is a steady encroachment on these boggy areas, and small stock, especially sheep, graze over the bogs, cutting into the plant cover of the surface, even starting erosion of the peat itself. As a result of this pressure, soil erosion has become a major problem in this country.

AGRICULTURE

Maize is the principle crop grown in this zone, followed by sorghums and wheat.

Tobacco is widely grown mainly for local consumption. Agricultural production is very subject to crop failures on account of extended droughts and this creates hardships to the indigenous population.

Livestock production

This is the most important sheep producing region of South Africa. Sheep meat is mainly consumed locally. High quality wool is produced from Merino sheep and this is Lesotho's most valuable export.

Great losses of stock occur during droughts, and to alleviate this many farms now have their own water supply.

The Bantus regard their cattle more as tokens of wealth than as food producers, so although cattle do exist in fair numbers, they are not adequately utilized. Goats are also raised for local consumption. Because of overstocking with livestock, the range in the Bantustans is generally in a poor condition.

Forestry

There are extensive plantations here of pines, wattles (*Acacia* spp.) and *Eucalyptus*, which are reasonably productive. Their wood is mainly used locally for various purposes. These plantations also serve an important function as protection forests.

Wildlife

In Lesotho the Bantu population is high, the land is overgrazed, and wildlife numbers have become seriously depressed. On the other hand, in the Orange Free State the farming community has become aware of the responsibilities of conserving the indigenous fauna and in areas of marginal agricultural value game ranching may well prove to be of greater financial importance than ranching, using domestic stock only. Many European ranchers now resort to game ranching as part of their operations and maintain stocks of blesbok (*Damaliscus dorcas*), springbok (*Antidorcas marsupialis*), black wildebeest or white-tailed gnu (*Connochaetes gnou*), grey rhebuck (*Pelea capreolus*) and mountain reedbuck (*Redunca fulvorufula*) (Von Richter *et al.*, 1972). A number of species not known to have occurred here in historical times have been introduced such as Hartmann's mountain zebra, waterbuck, red lechwe, bontebok and impala. These species gradually increase in numbers.

Prospects

Various kinds of anti-erosion schemes should be continued or implemented vigorously, including contour furrowing, grass stripping and terracing, if further deterioration of the land is going to be stopped.

The scheme for restoration of mountain pastures, initiated some years ago by the colonial administration in Lesotho, should also be continued.

The numbers or kinds of livestock which may for any specific period be grazed on any definite portion of land should be drastically reduced in heavily overgrazed areas.

The Kalaharian zone

Geographic distribution

The Kalaharian zone is situated on a plateau generally more than 1,000 m above sea level in Botswana, eastern South West Africa and north western Transvaal, stretching from the Okavango River in the north to close to the Orange river in the Republic of South Africa in the south (Fig. 4). It is flat or gently undulating, with sand dunes more frequently occurring in the south west. The true Kalahari is a huge sand-filled basin.

Climate

The zone is semi-arid. Rainfall is erratic, confined mainly to the period November to April and decreasing from about 500 mm annually in the north west to about 200 mm in the south west. From 22° latitude southward the zone is more truly desert, receiving only little and unreliable rainfall. The summer climate is dry and continental.

Vegetation

Scarcity of water and poor soils are the main limiting factors to vegetative growth. Surface water is available in only a few areas for a short time after the rains, when it collects in 'pans' or shallow depressions. These pans play a very important role in the ecology of the area.

The vegetation varies from open grassland to bush and tree savanna in which various species of *Acacia*, *Commiphora* and *Boscia albitrunca* predominate. The northern part of the zone has neither the climate nor the vegetation characteristics of an arid area and supports stands of trees and grasses, and luxuriant growths of ephemerals following the rains. The southern, most arid part of the zone supports scattered small trees (chiefly *Acacia* species), shrubs and, in occasional wet years, summer grasses. Here gourds are prominent, particularly *Citrullus vulgaris*. These plants are so abundant that they have made it possible for bushmen to live at considerable distances from water-holes. Succulent plants make up a large portion of the vegetation. The Kalahari is not a true desert, for there is a partial covering of grass.

Effects of Human Influence

The human population density is low, under one person per square mile over large areas. Settlements tend to be clustered near drilled wells (also called boreholes), wells and pans (waterholes), leaving large areas between these virtually uninhabited. Despite the low population density the vegetation has been highly modified in many parts of the zone by past land use practices, especially harmful burning habits and over-stocking of domestic animals.

Vegetation modification is most noticeable in the vicinity of water sources and often extends up to ten miles from such watering points. Deterioration of the perennial grass cover, lowering of the water table and the drying up of perennial springs in some of the pans are all linked to the human activities of the past and have resulted in destruction of habitats of a number of large mammal species (Von Richter, 1970).

Although this zone has been inhabited for at least half a million years, it was not until 1700 A.D. that a fairly advanced system of agriculture appeared and the region was settled in any numbers.

Great changes have taken place in this zone during the recent past. Before and around 1840 the headwaters of both the Molopo and Nosob rivers were flanked by much more luxuriant vegetation than exists today, although it is doubtful whether the year round flow in these rivers was significantly greater. The Limpopo drainage was well wooded, especially along the rivers, and the northern tributaries were infested by tsetse fly. Between some of the rivers, and especially around their headwaters, there were extensive open plains.

The tsetse fly has now been eliminated and scrub has invaded many of the open grassland areas. Rivers which once had perennial pools, at least as far west as the north-south railway line, now have less water in the dry season and many of the springs in the west of Botswana have disappeared, including even formerly dependable water sources where major villages were established. The southern Kalahari was much less wooded than at present and occasional springs existed which were used by bushmen and game.

Generally, these unsatisfactory trends have taken place too rapidly to be blamed on long term phenomena such as changes in climate. Many of the most degraded areas have been those in which there were large numbers of people and their stock at some point during the last 130 years or so. Thus within a relatively brief time span there have been very significant changes, mostly detrimental, to what once were the most productive areas (Campbell & Child, 1971).

Rangelands

In general, fire has had a detrimental effect on range conditions, and the effects of improper burning practices often have been less localized and have taken longer to manifest themselves than those caused by severe overgrazing.

Range deterioration in this zone is often characterized by the drying up of perennial sources of surface water, by a reduction in the vigour of perennial grassland and by encroachment of woody plants.

Grassland has been replaced by scrubland or in isolated areas by denuded soils, including mobile sand dunes, and these in turn have led to important changes in the wild fauna.

Animal Husbandry

The economy is still largely dependent on livestock production. Most livestock is free ranging and centered around permanent water, wells or boreholes beyond the limits of the tsetse fly. In the absence of proper control of stocking rates and range management, rapid range depletion continues to take place.

In North Eastern Botswana the introduction of borehole wells has led to increases in the number of cattle which has led to depletion of the grass and an acceleration of the gradual drying up of the permanent streams.

Wildlife

This zone is rich in wildlife, partly due to the sparse human populations and also due to the intrinsic productivity of the area resulting from the complexity of contrasting habitat types, often occurring close together (Child, 1970).

The most striking features of the wildlife of the zone include a remarkable diversity, including 46 species of mammals that are jackal-sized or larger; the ability of several species in the Kalahari to survive without surface water for most of the year; the mobility of some of those species like springbok who cover vast areas and the large herds in which they may occur (Child, 1970).

Although this zone continues to support high wildlife populations, hunting may have affected the disappearance or reduction of several species. However, a more likely reason for population changes appears to be changes induced in the vegetation as a result of overstocking of livestock and increase in fire incidence.

The historical evidence reviewed by Campbell & Child, 1971, shows that the ranges and numbers of some species like roan, sable and tsessebe have declined but that others have increased, the wildebeest and hartebeest in particular, along with elephant, buffalo, impala and springbok. The principal reason for the decrease of some species is the impoverishment of perennial grassland. It is important to note that increases in wild ungulate populations resulting from changes in the habitat induced by man and his domestic stock are liable to be of rather short duration. Certain springbok and wildebeest populations have apparently already declined following an eruptive peak (Child, 1971). Giraffe have disappeared from most of their former range, and eland have declined throughout theirs. Large predators including the lion, leopard and cheetah are still present in fair numbers.

Deterioration of wildlife habitats is also linked with expanding human activity, including increasing numbers of boreholes and wells for domestic stock without adequate control of grazing practices. This has a detrimental influence on wildlife which depends on the available water to meet their essential requirements.

Subsistence hunting is a traditional form of land use for many people. Almost 60 per cent of the protein consumed within Botswana is derived from wild animals (Von Richter, 1970). Hunting supplies the rural tribesman not only with essential proteins but also with a not insignificant cash income. Since the early 1960's wildlife has become an important source of revenue and has contributed more and more significantly to the national economy of Botswana. The increasing demand for wild animal skins and the rising prices coupled with it, has stimulated the villagers to hunt more regularly and offer the skins for sale to the local traders. An industry based on hunting by rural people, adequately managed and controlled, meets the basic requirements of a sound long term venture without endangering the ecological balance, provided it is protected from competing forms of land use such as cattle ranching. Safari hunting has also increased considerably.

Prospects

The human population is increasing rapidly and already Botswana is very dependent on outside aid to feed its people. The implications are frightening when it is recalled that large numbers of people did not settle here until about 150 years ago, and that cattle numbers were not significant until the last 70 years or so. Somehow, preferably by reducing stocking rates, the damaging effects of cattle ranching must be stopped or at least be alleviated. Very high priority must be placed on halting the destruction of wildlife habitats by man and his stock. This is necessary not only for the sake of wildlife, but also to safeguard the future of livestock industry. There is an obvious need to develop land use patterns which are suitable to conditions in this zone. The great value of wildlife demands that its proper management should form an integral part of land use in large sections of the country (Child, 1971). Moves are now under way to exploit the water resources of the Okavanga Swamp in northern Botswana and elsewhere. Hopefully the impact on the ecology of the swamp will be carefully studied before any development takes place. Increasing use will be made of pans for human settlements, resulting in damaging effects on wildlife and the surrounding habitats.

The Karroo-Namaqualian zone

Geographic distribution

This zone consists of the Great Karroo and Little Karroo plains which are situated on the Karroo plateau and encircled by mountains. These plains are generally more than 1000 m above sea level and slope gently northwards towards the Orange river. Namaqualand occupies the driest, western part of the plateau. The zone also includes a relatively narrow strip of arid land along the west coast stretching from southern Angola across South West Africa into the Cape Province of South Africa. The most arid core of this strip is the Namib desert (fig. 4).

Climate

Under the influence of the cold Benguela Ocean Current the west coast receives but scanty rainfall – being less than 80 mm per year – while the Namib desert receives less than 50 mm. As a result, desert conditions prevail in the coastal belt. The Karroo plateau receives occasional winter rains, but precipitation averages less than 25 cm and the seasonal nature of the distribution of rainfall is just as limiting as the scanty amount. The Great Karroo forms the most arid part of the plateau and can be considered as a a southern extension of the Kalahari. The plateau suffers from extremes of temperature due to its continental climate with severe frosts in winter and dry heat in summer.

Vegetation

Since the first white settlement (1652), range conditions in this zone have gradually deteriorated. The original vegetation on the Karroo plateau, and to a lesser extend the coastal strip, was mainly grassland. The flora of the Little and Great Karroo is characterized by dwarf trees, shrubs and succulents, the latter element being stronger in the Little Karroo. Large stretches of this zone are treeless, covered only by very scattered shrubs of small size. For the rest bare soil remains. Most plants are found among rocky outcrops and in dry river beds. As water storage is of prime importance to tide the plants over the hot dry summer season, many succulents and geophytes have evolved. There are also bulbous plants, acacias and tamarisks. Other plants survive the hot, dry summers in the form of seeds.

In the absence of rain, sea fogs and dew are of importance in sustaining the little, mainly succulent, vegetation that survives in the Namib desert. The Namaqualand vegetation consists of scattered small shrubs and succulent plants.

Effects of human influence

Much of this zone has been severely overgrazed by sheep and goats and in fact, it offers one of the most striking examples of the sensitivity of African vegetation to browsing, grazing and trampling by livestock. Acocks (1963) has mapped the degradation of the vegetation that has taken place during the occupation by African and European herdsmen in recent centuries. The species composition of the Karroo veld has changed: perennial grasses have become scarce or have vanished; succulents have become more important and the vegetation of valleys has disappeared (Acocks, 1964). Within living memory, sheep and goats have converted much palatable, nutritious, xerophytic and partly succulent subscrub and grass vegetation on the Karroo to bare ground, alternating with dense communities of species either little relicts or wholly rejected (Phillips, 1959).

The Little Karroo, the first to be subjected to settlement and grazing, has suffered particularly severely from overgrazing. Very little of the original top soil is left and few, if any, relicts of the climax vegetation survive.

Soil erosion and the encroachment of inedible and poisonous plants is widespread. Invasion by Karroo bushes does not check erosion as these are able to exist under very poor conditions and do not have a growth habit which will check erosion.

Parts of the high-altitude Karroo (between 1,000 and 1,150 m) have become desert because of the grazing out of all Karroo bush (Acocks, 1964). Over many thousands of acres of pasture land in the Karroo so much soil has been washed away that the crowns of the bushes stand 15 cm or more above the soil instead of resting on it, and the seeds from the remaining bushes cannot establish themselves (Hailey, 1957).

In the report of the Long-Term Agricultural Policy Commission (1949) great stress was laid on water as a limiting factor in the progress of rehabilitation, as indicated in the following extract: 'What has been proved, on the other hand, is that the land has already become more arid, as indicated by diminution in the flow or drying up of springs of pre-European occupation times, by intermittent streams and rivers becoming more intermittent, by sinking of the water table in wells and boreholes or the drying up thereof altogether, that the character of the vegetation over extensive regions in many parts is becoming gradually more xerophytic and that secondary plant communities are replacing the primary and becoming established therein; consequently the land is becoming more bare and erosion by wind and water is increasing extensively and intensively.... On its own observations, the Commission is forced to the conclusion that such conditions have been induced by interference'.

Animal husbandry

In this very thinly populated zone the people often suffer greatly from drought. The vegetation supports very few sheep and goats. The Great Karroo is the homeland of the South African merino sheep. Karakul sheep are being raised successfully in the Namib desert.

Agriculture

Immediately below the Great Escarpment small streams are utilized in a narrow ribbon of cultivation with wheat and lucerne as typical crops' The irrigated belt along the Sundays river has a flourishing industry of citrus fruits.

Wildlife

Since early settlement days wildlife populations have been much depleted because of the habitat destruction and overhunting. Even if the hunting pressure can be curtailed, wildlife populations will continue to be low as a result of poor range conditions.

Prospects

The improvement of range management and grazing are the main problems, and a major obstacle is the shortage of water supply. Successful improvements can only be obtained if the stocking rate of livestock is drastically reduced for a number of years. If the range of the Karroo is to be improved, drastic conservation measures must be undertaken including the establishment of a close sward-forming grass cover, the application of proven grazing systems which will ensure that the grass will not again be overgrazed, and the construction of weirs and embankments (Acocks, 1964). More windbreaks should be planted to counter wind erosion.

The Cape zone

Geographic distribution

This is by far the smallest ecological zone and covers a relatively narrow strip along the southwestern tip of Africa, consisting of a coastal belt and its adjacent mountain slopes. Scattered mountains occur also in the coastal belt. It is bounded by the dry Karroo-Namaqualian zone to the north and the Transvalian zone to the east (see fig. 4).

Climate

This zone has a mediterranean climate, namely high precipitation during the southern winter (October to April), and dry summer months. The coastal belt receives a relatively high rainfall (65 cm), but the mountains receive as much as 150 cm. Frost is rare in the lowlands, but snow frequently falls on the mountains.

Vegetation

The zone is typified by a great wealth of species of plants, many of them occurring nowhere else, such as the Proteaceae. The richest floral region is situated in the coastal belt: about 2,500 species of higher plants grow on or near the Cape of Good Hope. The mediterranean type of vegetation called 'maquis' or 'macchia' consists mainly of thickets of wiry shrubs with hard, green leaves.

Indigenous forests are confined mainly to the constant rainfall region along the coast in the eastern part of the zone, they are also found on the moist southerly slopes of the mountains. Most tree species are evergreen.

Effects of human influence

As this was the first part of South Africa to be colonised by Europeans and is now relatively densely populated, much environmental damage has been done. Acocks (1963) inferred that in all probability until recently a continuous belt of forest and scrub-forest 80–230 km wide extended round the entire east and south coast of what is now the Republic of South Africa. This has been so generally removed that few forested areas remain. The indigenous vegetation has also largely disappeared from more arid parts of the zone. In the south western Cape Province the vegetation is deteriorating rapidly mainly due to burning and grazing practices and the steady spread of Australian species of *Acacia* and *Hakea* presents a serious problem. In the arid and semi-arid areas of the north west Cape thousands of square km of land have gravely deteriorated. Extensive areas of the Cape Province which once were highly productive wheat lands have been forced out of cultivation (Bennett, 1945).

Agriculture

Mixed farming has been carried on by Europeans since early settlement and practically all cultivable land is now tilled. Maize is produced as a subsistence crop. Viticulture is confined commercially mainly to the south west Cape and wheat is grown as a winter crop in the same region. The manufacture of wines and brandies is carried out on a considerable scale. Various kinds of fruits are grown in extensive orchards and give rise to many industries, including the production of fresh, dried and tinned fruit for export.

Animal husbandry

Livestock production, consisting of the raising of beef and dairy cattle, goats and sheep, is not a major industry in this zone. However, much of the eastern part of this zone is devoted to cattle production.

Wildlife

As a result of intensive agricultural development since the start of European settlement in 1652, wildlife and its habitats has become drastically reduced. One species that has been seriously endangered with extinction is the beautiful bontebok (*Damaliscus dorcas*). Careful protection of this species since 1930 has resulted in the present population of more than eight hundred. Another rare species now virtually restricted to the Cradock National Park and a nature reserve is the Cape mountain zebra (*Equus zebra*). Most of the remaining wildlife populations survive in well-protected reserves.

Prospects

There is a continuing need for range improvement, including the control of grazing and the reduction of overstocking. This should take precedence over dam construction or similar engineering works.

III. MAN AS AN ENVIRONMENTAL AGENT

Primitive man's influence on the environment

Since the last ice age man has increasingly modified his surroundings and, in fact, the whole history of *Homo sapiens* has been a struggle for dominance over the natural environment. With the passage of time man has become even more capable of controlling nature, or at least of utilizing certain aspects of it to his own advantage. In the initial stages of food gathering, such as is still carried out by the pigmies, man modified his environment little, or not at all. Stage by stage, his impact on his environment has increased, but only during the last centuries has this become disastrous. As a result, some environmental restraints on his population have been removed, but this has been at the expense of other organisms and has brought out depletion and exhaustion of natural resources.

The researches of LEAKEY and his cooperators indicate that humanoid beings existed in Africa approximately two and a half million years ago, the oldest known record of early man. Initially his effect on the environment must have been slight as subsistance activities were restricted to the gathering of fruit, roots and grains, but with the development of primitive weapons for hunting and fishing this effect increased, although it was still not very pronounced as human populations were still too low to exert much influence.

Nevertheless, MARTIN (1971) presents evidence to substantiate that a major episode of Pleistocene extinction antedates the late Würm (40,000–50,000 B.P.). Although Africa's fossil fauna is far from adequately known, he estimates that roughly fifty genera disappeared during the Pleistocene. The living genera of African big game represent only 70 per cent of the Middle-Pleistocene complement. Thus, despite its extraordinary diversity, the living African fauna must be regarded as depauperate. Late-Pleistocene extinction in Africa long precedes that in the Americas and Australia, as could be expected in view of man's early evolution in Africa. Martin observes that extinction on the continent seems to coincide with the maximum development of the most advanced early Stone Age hunting cultures.

PRIMITIVE MAN'S USE OF FIRE

Man first acquired the use of fire probably well over 350,000 years ago (CLARK, 1959). No doubt in prehistoric times grass fires occurred naturally from time to time, but man's burning would have tended increasingly to change the vegetation because the same area was burned more frequently. Although relatively few fires may have been set, they probably burned extensive areas if started during the driest time of year and therefore may already have had a considerable influence on the environment.

When he began to live in caves or rock shelters man must have learned to make fires to provide warmth and this ability enabled him to penetrate altitudes and areas which were formerly too cold for habitation. As he moved into forested zones, probably this use of fire led to some unintential destruction of forests. It was only in the very late stages of prehistoric times when he had established a limited amount of agriculture that fire was used to burn excessive crop residues as well as to clear away old, dead grass and to encourage a new growth from the underground parts for the grazing animals. In the later stages of his hunting existence man probably used fire to drive herds of game so that these animals could be slaughtered easily. The existence of most open grassland lacking woody growth of any significance can be attributed to man's influence, although some natural grasslands exist for edaphic reasons, particularly in the arid and semi-arid zones.

Primitive man as a cultivator

Environmental damage accelerated when man became a cultivator. It is not completely certain when this occurred in Africa and no doubt there must have been a very gradual changeover, but it must have started over 7,000 years ago, at least in the Nile delta as is evidenced by archaeological information. Cultivators spread from Egypt and Ethiopia to the south and the west to reach many of the other people of Africa. With only primitive tools naturally man would have cultivated first those areas which were most favourable such as open grasslands and alluvial river valleys, and produced crops such as barley and wheat. As a consequence of ploughing and harvesting, the soil soon became exposed to the elements and this resulted in loss of water through increased evaporation as well as loss of soil humus. After developing metal ploughs and axes, man extended his agricultural activities even further into forested areas, where the soil was richer in humus and rainfall more ample.

Production was of course correspondingly greater, but soil depletion was often also more extensive.

Primitive animal husbandry

The next development in the disturbance of the environment came with the domestication of stock which began in the temperate parts of Europe and Asia and in the Near East and later spread to the Nile Delta. No doubt, stock-keeping spread southward from the Neolithic peoples of the Mediterranean and the Sahara, but since in Cyrenaica, for example, domestic stock were first kept less than 7,000 years ago, these animals were presumably not an important factor in the tropics until after this date (Moreau, 1966). In fact, it seems that domestic stock has been present in the tropical parts of Africa in large enough numbers to damage the habitat only in more recent times, say the last 2,000 years. The transition to agriculture and stock-breeding which made possible the development of settlements and villages spread but slowly down the Nile Valley. The penetration of the Bantu and his livestock into Rhodesia has occurred more than 1,200 years ago (West, 1972). Although the impact of domestic stock on the range in central and southern Africa is of relatively short duration, it has occurred over a long enough period to cause considerable localized damage to the environment by continuous overgrazing and overbrowsing pressures.

Implications of the impact of primitive man's influence

A most conspicuous characteristic of man is that he destroys habitats to suit his own expanding needs and in doing so affects both plants and animals. During recorded history he has been instrumental in causing the complete extermination of a number of mammals and birds in Africa, particularly so in North Africa (see p. 137). Little detailed information is available on the elimination of certain plant species (see p. 140), but it is certain that many of them have also suffered through his activities. Through the centuries man has become increasingly powerful in modifying the plant and animal world surrounding him and history shows the long and diverse series of steps by which he achieved ecological dominance. He has intervened to increase and decrease species of plants and animals, to expel or exterminate and to introduce and to modify organic entities.

Although in most parts of Africa the activities of primitive man have been replaced by those of more advanced husbandry, in the more remote parts of the continent primitive tribes still continue traditional practices in a more or less unmodified manner. Such people live wholly by hunting, fishing and foodgathering. Some bushmen, pygmies, pygmoid, Batwa, Wanderobo and negroid hunting people known collectively as wata still lead the life of the Old Stone Age (Allan, 1965). The bushman, who at one time ranged and hunted over much of south and south-central Africa, from the Cape to Rhodesia and from Angola to Mozambique may be the most numerous of the survivors, although now they are restricted mainly to Botswana. There is inadequate information about the land requirements of present day hunting men of Africa, but it is generally agreed that they need very extensive areas for their food gathering.

The necessity of preserving a balance between population size and available land area must have presented itself early in the human story and we may therefore suppose an ancient and practical origin for certain customs and devices of savagery that tended to maintain this balance. Primitive tribes occupied 'hunting or food gathering territories' which met their basic requirements and these territories were mutually recognized by adjacent tribes with the result that the food gathering activities did not exceed the carrying capacity of the territory used. Even today a concept of 'land tenure' of exclusive rights over lands or its natural products is found in many of the surviving cultures (ALLAN, 1965).

In present day Africa practically all the cultivating peoples still supplement their diet by gathering food, by fishing where they may, and by hunting where there are still beasts to hunt. Indeed in many tribal systems, an important part of village subsistence is obtained by full utilization of the surrounding bush, woodland or field.

Here and there one still comes across some curious remnant of the old economy. In north-eastern Ghana an indigenous grass, *Dactyloctenium aegyptiacum* is still laboriously collected for food, while in Sierra Leone, Guinea and parts of Nigeria another grass, *Digitaria exilis*, which has not been found in the wild state, is cultivated as a grain. 'At least until very recently, the Valley Tonga (of Zambia) also harvested and sometimes even stored wild grasses, the seeds of which were boiled as grain. These probably played an important part in alleviating the recurrent famines of the past' (ALLAN, 1965). Fine-seeded annual grasses such as Teff (*Eragrostis abyssinica*) are still staple bread-grains of Ethiopia (SEMPLE, 1971). In summary, it can be said that although primitive man definitely had a modifying influence on his environment, it was not sufficient to seriously affect its ultimate productivity, besides which the numbers in relation to land size were small and large tracts of land were thus left in a virtually untouched condition.

Modern Man's Influence on the Environment

Only a relatively small fraction of the African continent has a human population density of more than 10 per km^2 (Demographic Yearbook 1969), yet over most of its vast surface the influence of modern man on the vegetation has been considerable.

In the humid, tropical zone the exposure of soil to heat and intensive rainfall after clearance of the forest renders maintenance of soil fertility impossible. Deforestation, irrigation, the introduction of exotic plants and animals and the large-scale use of weed-killers, to mention but a few examples of changes wrought by man, have profoundly transformed the tropical landscape into a less diversified and much deteriorated ecosystem. The arid zones have been ravaged by fires and overgrazing and have lost much of their productivity due to these human influences. In both zones the microclimate has been affected adversely due to denudation of the vegetation.

The appearance of great urban communities which brought with them a more sophisticated technology is the most recent development affecting the African environment. Although large metropolitan centres are relatively few, their impact on surrounding rural communities is considerable and spreading constantly.

In our technological era two developments have influenced nature for the worse. Firstly, many habitats of scientific interest have been greatly reduced in extent or sometimes virtually obliterated. Many wetlands have been drained and are now used for intensive agriculture, and many dry lands have been ploughed up or replanted into man-made forests. Secondly, many wild animals and plants which were formerly harmless or of little concern to the farmer have now come to be regarded as pests or weeds and are ruthlessly eliminated.

Today many means of eradicating unwanted animals or plants exist but unfortunately since few such means, whether manipulative or chemical, are wholly selective, many innocuous species are eradicated along with the pests.

As the millions of human mouths increase, the demands on agricultural production rise even higher. The ultimate consequence of man's influence on his environment is predictable: ever increasing numbers of plants and animals, with the exception of those which have been domesticated or are of proven value, are likely to be considered pest or weed, and as such will be increasingly threatened by the axe or reduced by chemical treatment and we will one day be faced with a severely impoverished environment. Table 1

Table 1. The Impact of Man on the African environment from primitive times until the present

Stage of Land Use	*Sources of Plant and Animal Foods Used*
1. Primitive hunters and collectors	Wild plants, including roots and tubers, leaves, shoots, seeds, fruits, nuts and fungi. Wild animals, including mammals, birds, lizards, insects and worms. Collection of honey.
2. Shifting cultivation, with regenerating secondary forest as fallow	Wild plants and cultivated cereals and legumes. Wildlife of successional plant communities. Some domestic livestock.
3. Shifting cultivation, with inadequate regeneration of secondary vegetation; soil fertility tending to fall	Mainly grain crops, limited vegetables. Contribution from wild flora less significant. Contribution from wild fauna less substantial. More domestic stock, scavenging.
4. Settled dryland cropping, with continued use of shifting cultivation mainly on infertile soils.	Mainly cereals and grain crops. Not much use of wild flora and fauna. Increasing use of livestock.
5. Grazing and browsing of livestock. Communities nomadic or settled; some cultivated land.	Mainly use of livestock as a source of food, but some use of grain, legumes and vegetables.
6. Irrigated, plus settled dryland, plus shifting cultivation.	Mixture of several dryland cereals, wet rice and root crops. Legumes and vegetables. Much use of livestock. Only small wildlife and insects hunted.
7. Mixed farming: agricultural crops and livestock. Irrigation whenever possible.	Major cereals, grain legumes, fodder crops, cash crops. Integration of agricultural production and animal husbandry on an intensive basis. Wildlife use very little.
8. Plantation agriculture	Very intensive production of cash crops and tree crops mainly for export. Labourers keep various types of livestock and cultivate small private gardens.

shows the various stages of land use from primitive man until the present day and the various sources of plant and animal foods that were utilized. As all these stages are still practised somewhere in Africa no time scale has been added. The use of wild animals and plants decreases as the intensity of land use increases.

Implications of the human population explosion

No one who realizes that global resources do have finite limits can doubt any longer that control of the human world population is one of the major problems facing mankind. Nevertheless the urban populations are increasing at about twice the rate of overall population growth, thus justifying the need for the shifting of the primary urgency towards the control of the rate of urbanization and the planning of both urban society and environment in considerable detail.

The gross multiplication rate natural for man is very low by comparison with most animals. However, because of even primitive man's relatively advanced practices of protecting and providing for his family and himself, a fairly high percentage of young survived and reproduced. However, modern man's scientific and technological revolution has more than tripled the probability that a human being born in our time will survive through the age of reproduction and it has therewith tended to triple man's net rate of multiplication.

It has been estimated that the population density of paleolithic man was around 2,5 km^2 per person. Before the industrial revolution, that is the middle of the eighteenth century, the world population was 700 million.

As a result of improved chances of infant survival and decreased mortality this number passed one thousand million in 1860, two thousand million in 1930 and 3.5 thousand million in 1971. We can expect 4 thousand million in 1975 and 6.5 thousand million by the year 2000.

The great problem resulting from these rapid population increases is that as many as two-thirds of the world's present population is suffering from malnutrition to a greater or lesser extent and that the threat of large-scale famine is still with us despite the 'green revolution' and other nutritional advances.

No one knows exactly how many people Africa can support because no accurate censuses have been made in many countries. However, large areas of tropical Africa appear to be underpopulated in comparison to other parts of the world, certainly by European standards. The population density is also not comparable to the densely populated parts of S.E. Asia. Nevertheless, people live in almost all habitats, with the exception of the desert, drier savanna and high mountain altitudes. Many areas appear extremely sparsely populated in relation to the availability of cultivable land. In such areas a low population density is a definite handicap in the progress towards productive agriculture because of the inadequate numbers available to work the land. Many savanna lands such as most of the Central African Republic, Zambia, Eastern Angola and rainforest areas such as parts of the Congo basin and Gabon are all examples of such population sparsity. The African population is now growing at more than 2.8 per cent a year and may increase even faster in the years to come, due to better public health measures and disease control, together with the general attack on African poverty and malnutrition sponsored by foreign aid programmes. According to UN population data the rate of population increase over the next decade ('71–'81) is expected to range from around 1.4 per cent per annum in Angola and Gabon to 3.4 per cent in the Ivory Coast. A rate of increase between 2.5 and 3.0 per cent is expected in such populous countries as Nigeria, Morocco and Zaïre. Figures published by the UN in 1970 show that Africa had a population of 303 million in 1965. The projected population figures for 1970, 1980, 1990 and 2000 are 344 million, 456 million, 615 million and 817 million (World Population Prospects, 1965–2000, as assessed in 1968 – UN Population Division, Working Paper ESA/P/WP 37, December 1970).

Low, or at best, moderate population densities are typical of the Miombo Woodlands, suggesting that most of these have a rather low carrying capacity for sustained agriculture. In the double-rainfall zone bordering the equator, between the arid zone and the equatorial rain forest, there are regions of a high carrying capacity and correspondingly high population densities. Included in this category are such regions as Kikuyuland, Ruanda-Burundi, Kigezi and Teso (Uganda) and Sukumaland (Tanzania).

The number of persons per square kilometer varies widely in different parts of Africa:
Western Africa – 18 persons per square km
Eastern Africa – 15 persons per square km
North Africa – 5 persons per square km
South Africa – 8 persons per square km
(Demographic Yearbook 1969. UN Pub. 1970).

Although these data might give the impression that Africa is thinly populated and has much surplus agricultural land, in actual fact Africa's reserve of suitable land for further agricultural development is not great and the bulk of the necessary increased food production will in the future have to come from much the same land as now. Moreover, the overall population densities given for the continent conceal the very high population densities of well over 1000 per sq.km in many areas, such as the highlands of Ruanda-Burundi, Kenya, Tanzania and Malawi.

The urban population growth in African countries can be regarded as being the highest in the world. The towns and cities are drawing largely on the rural population (Urban and Rural Population: Individual Countries 1950–1985, and Regions and Major Areas 1950–2000 – U.N. Populations Division, Working Paper No ESA/P/WP 33 1970).

The migration from the farming communities into the cities is largely an attempted escape from hunger or insufficient employment. An example of an overpopulated rural area is the Kabale district, Uganda, where all the usual symptoms of overpopulation of the environment are evident: almost continuous cultivation and consequent soil degradation, sub-division and excessive fragmentation of land.

Increases in population, resulting from considerable improvements in health care since the 1950's, have caused the increasing use of land for animal and plant production, and this high growth rate is contributing considerably to the difficult task of economic and social development. Governments will find it extremely difficult, if not impossible, to improve social services and provide sufficient numbers of health and education facilities for their rapidly growing populations.

The fact has to be faced that in the main problem areas of Africa – and these are, in general, the areas of highest fertility and most favourable climate – there just isn't enough land to support the present numbers of livestock, nor can the existing land be improved sufficiently to support ever-increasing numbers. As a result, conditions of life are bound to worsen for every individual. The problem of redistribution and resettlement of people from overpopulated areas is far more difficult than it might appear. Although there are extensive areas which are very sparsely inhabited or even in parts uninhabited, these are mainly drought lands, swamp regions or areas subject to very heavy precipitation throughout the year, where reclamation calls for very heavy and in most cases uneconomic capital expenditure. The development problems of the very poor soils which are so common in Africa have received little attention, and it seems improbable that under prevailing conditions any economic system capable of maintaining considerable population densities can be devised for them. A controversy is raging among demographers, agriculturalists and ecologists about the extent to which the present human population explosion can be allowed to continue and what 'optimum' populations are acceptable for the different environments. No one really knows the answer to the second point because little relevant research has been undertaken so far. Also, opinions differ greatly over what constitutes an optimum density. On the basis of personal observations I would say that there are already many areas on the continent where human populations have exceeded the capacity of the land to sustain them. These include not only areas where human populations are most dense, as in parts of East and West Africa, but also many areas where even low human densities make too great a demand on the land, such as in the Sahelian and Karroo-Namaqualian zone, where the productivity of the land is extremely low. The areas that are not now affected by excessive human populations from an ecological point of view are rather few and far between and include parts of Angola, Gabon, Zaïre and Zambia.

It seems that man's need for space and a certain amount of solitude is real. Even if technology could produce enough synthetic food for all, overcrowding produced by ever-increasing populations is likely to have disastrous social and ecological consequences. For this reason and those given previously, further settlement of the now sparcely populated regions should be discouraged. Efforts should be dedicated preferably to a better use of the productive lands through the application of improved techniques and the education of the rural populations.

Effects of Fire on the Environment

Fire has been and still is the principal clearing tool of primitive people. Large parts of Africa have of course been burned regularly since time immemorial but are being more so recently due to the widespread use of matches, with the consequence that the effects have become more pronounced. The setting of fires by man has resulted in deep and lasting modifications of the African vegetation and it is because these activities have exerted such a profound influence for so long that fire has become an important environmental factor almost equal in effect to that of topography or climate.

Certain fire-adapted communities have developed through evolutionary processes where plants and animals have become adapted to periodic burning as a normal feature of the environment. Burning is widely used for the so-called improvement of grassland: the litter is removed and a fresh flush of vegetation results which has a higher protein and mineral content, and is therefore more nutritious. This attracts animals, but the actual benefits which they derive from it are somewhat doubtful, because the abrupt switch from a relatively poor fodder to this highly nutritious one probably causes intense rumen microbial activity, resulting in scouring and sometimes even death amongst certain herbivorous mammals.

Unfortunately, grass burning is one of the most widespread of malpractices in Africa due to its frequency and method of application. One result of this is increased erosion which is caused by the burning of the humus in the upper soil layer, which is also detrimental to soil structure. Another result is the modification of the vegetation, harmful to grazing animals because it favours the spread of fire-resistant bush and less nutritious grasses. The more nutritious, softer grasses are the most susceptible to fire damage and consequently the more unpalatable grasses increase in abundance. One of the more serious effects on animal life is the removal of essential cover which not only affects the available food supply, but also can be even more disastrous when burning gets out of control and some animals themselves are killed in consequence.

Grazing and burning are intimately interrelated, since heavy grazing removes fuel, thus preventing fires and allowing growth of the ungrazed woody vegetation, ultimately producing enough fuel to create new fire hazards.

Fire is an important factor in maintaining grasslands in their present form, and the exclusion of fire leads to increased establishment of trees and shrubs and permits vegetational succession leading gradually towards some type of wooded plant community. Frequent fires are not hot enough to kill shrubs and young trees because there is an inadequate accumulation of fuel. This survival of shrubby vegetation ultimately leads to bush encroachment and thickening of tree cover. Thus, although fires do injure plant communities to a greater or lesser degree, they are not always detrimental and under the right circumstances desirable effects can be maximized. The effectiveness of fire in producing and maintaining grasslands is dependent on the season, on the intensity of the fire and on the degree of tolerance towards fire exhibited by the woody elements of the vegetation. Because the accumulation of dry plant material on the soil is little in more arid areas, fires are relatively infrequent, but if a fire does occur, recovery of the vegetation is extremely slow.

Fire and Soil Conservation

One of the effects of burning on the soil is that the parts of the grasses and herbs above ground are not returned to the soil to form valuable humus, and merely provide ash. Although this ash settles and will later be washed into the soil, some of the nutrients are lost in the burning (particularly nitrogen and sulphur), or will be lost before they can be taken up by plants. Fire also reduces the water retaining capacity of the soil and increases surface run-off, and that combined with the natural fall of the water content during the dry months causes the grass to be more subject to drought injury. Continuous use of fire causes xerophytization of rangeland and a lowering of the level of the ground water. The removal of the grass cover exposes the soil to danger from erosion when heavy rain falls, especially on steep hillsides. Absence of vegetation of course also increases the effects of wind erosion. Such effects can be readily seen almost anywhere in Africa.

Fire as a tool

Opinions vary about the use of fire in habitat management, but the majority of recent authors (DAUBENMIRE, 1968; PHILLIPS, 1965) consider fire a useful and cheap tool. The problem to be solved is when and how often burning should take place. It can safely be initiated at the start and the finish of each rainy season. Advantages of an early burn are, because of inadequate dry, combustible material that it doesn't get very hot and therefore doesn't damage the trees and shrubs too much but does reduce the possibility of subsequent fires. On the other hand, a disadvantage is that during the rest of the dry season the pasture doesn't produce any longer because the dry matter that would otherwise have been available has been burned off. The advantage of a late burn is that the fire is much hotter and this kills unwanted shrubby vegetation.

In addition, the underground parts of the plant that have not been killed sprout earlier, thus providing food, but if the rains do not then fall, heavy pasturing will weaken the grasses.

Carefully planned use of fire can provide grazing animals with a nutritive food source upon a sustained yield basis. The tracts or patches to be burned should be large enough to avoid excessive bunching and overgrazing by the animals. Preferably the burning rotation should not exceed three years.

Effects of agriculture on the environment

In this section special reference is made to the widespread effects of shifting cultivation and to some of the influence which colonial powers have had on agricultural practices, particularly with regard to cash crops.

A wide range of land use systems have evolved through the centuries which are often remarkably ingenious and adaptive to the environment's demands. Trapnell & Clothier (1937) have described some systems in Zambia. There for example, certain forest tribes rely for fertility on the ash of trees which they fell to make their gardens. Brushwood is generally hauled in from an area much greater than that cultivated in order to obtain the necessary supply of ash.

Traditional systems of land use

Even now the bulk of agricultural production still takes place in a traditional manner in which the principal tools of the peasant farmer remain the hoe, cutlass, axe and knife. In most areas attempts are still made to maintain the fertility of the land by rotating crops with bush fallow (see p. 115).

The nature of these systems, and of the social systems with which they were bound was rarely understood or even perceived by the suzerain peoples (Allan, 1965). Unfortunately, even now most technical aid programmes have an inadequate understanding of such systems which were highly labour-intensive and included elaborate anti-erosion measures such as terracing, as was, for example, carried out by early peoples in south western Uganda.

Modern systems of land use

'Modernising' African land management systems does not always serve the land well from a conservation point of view and when mechanization is introduced it often leaves people who would otherwise be employed tilling the land with no means of livelihood.

Many traditional systems of land use have degenerated or collapsed altogether under the impact of explosive population growth, cash cropping, loss of land, social disruption, labour migration and other changes brought about by European intervention.

Fortunately many of the continuing attempts to improve and develop African land-use have been at least partially and sometimes very successful.

In Northern Nigeria long and persistent efforts, first by the colonial government and after independence, by bilateral assistance, extending over more than thirty years have resulted in a not inconsiderable development of 'mixed farming'. Under this system the large cattle population feeds on crop residues during the dry season and their manure is accumulated. Cultivators who value manure as fertilizer make token payments to herdsmen or provide them with subsistence rations. Another extremely successful effort through multilateral and bilateral assistance in several parts of West Africa has been the growing of rice, mainly as an irrigated crop. Although originally instigated as an experimental effort by external assistance technicians, particularly the Nationalist Chinese, Africans have taken well to it. Further examples are the extremely productive palm and citrus plantations in North Africa. The 'green revolution' has also made its impact in Africa and highly productive hybrid varieties of maize are now grown in eastern and southern Africa.

The rehabilitation of the Kikuyu lands

Many resettlement and co-operative farming programmes have been tried throughout Africa, but unfortunately the results of most of them have fallen far below expectation.

One such programme at least achieved considerable initial success, that of rehabilitation of Kikuyu lands. In 1954 a five-year 'Plan for the Intensification of African Agriculture in Kenya was instigated by the UK Government following the Mau Mau uprising. The purpose of this plan was to change a deteriorating area into individual consolidated holdings. It was followed by subsequent programmes after Kenya's independence.

A main reason for the initial success of this scheme lies in the favourable Kikuyu environment. Its excellent soils and a climate conducive to grass growth of high feed value posed no very difficult technical problems, and experience gained in more temperate regions was more readily applicable here than elsewhere in Africa. Another advantage was the possibility of growing high-value cash crops. Thus, a comparatively small acreage would suffice to provide the minimum desirable standard of living for a family. The initial success of the programme is also a tribute to the Kikuyu people themselves, who are in general anxious to learn new methods of agriculture.

Land consolidation – the gathering of each family's scattered holdings into a single unit – progressed rapidly after initiation of the scheme. Consolidation was never imposed by compulsion. The success was probably largely due to the co-operation of tribal elders and other men of influence, and to the generally high quality and uniformity of the land. Consolidation was an essential preliminary to better land use, which was to be brought about by grass-arable rotations, the use of manure to build up fertility, soil conservation practices, and the planting of cash crops.

People who showed an aptitude for progressive farming by successful practice of elementary principles of good land-use became eligible for assistance in the form of farm planning.

The problem of finding suitable livestock for the improved farms was not an easy one, despite the excellence of the feed and the comparative ease with which grazing could be established. It was obvious that high-yielding stock would be required to make full use of potential grazing areas and to ensure an economic return from these grasslands. For this reason European breeds were introduced which fared well under the favourable environmental conditions.

Unfortunately this is not the end of the story. The birth rate among Kikuyu's is among the highest in Africa: the high population pressure which has prevailed during the last few decades has nullified many of the good results obtained in the initial settlement schemes, and throughout most of the Kikuyu lands signs of land degradation are becoming increasingly apparent. Soil erosion is visible everywhere and inadequate use is made of cover crops to reduce the effects of erosion. In addition, many protection forests have been invaded by settlers and much cultivation is done on slopes which should have remained under forest cover. HUMPHREY (1945) concluded that the problem posed by population densities on the Kikuyu lands, little short of 212 km², can be solved only by a drastic reduction of the agricultural population in addition to radical changes in agricultural practice.

It seems therefore that many of the good results obtained by the initial settlement scheme have since been done away with as a result of increasing population pressure. What has been said about the problem of the rehabilitation of the Kikuyu lands is clearly illustrative of what is likely to happen elsewhere in Africa where similar conditions prevail. Unfortunately there is not much room for optimism.

The initial success and the ultimate failures described here are only illustrative of similar developments elsewhere on the Continent.

If agricultural development projects, often heavily subsidized by multilateral and bilateral aid, ultimately prove unsuccessful simply because there are too many hungry mouths to feed, there seems to be little purpose served in the exercise. There is no need for taking a defeatist attitude in this regard, but at the same time if the matter of external aid to such projects is requested, the aid-granting agencies might well request a population control policy from the governments concerned before attempting to assist these governments with further rehabilitation schemes.

Variations in the cultivability of land

There is obviously a wide divergence between the proportion of land which can be cultivated by a people with advanced techniques and sufficient capital, and a people who possess neither of these resources, nor the skills to use them. In considering African systems of land use we must accept the limitations imposed by traditional implements, skills and customs, and adapt our own conception of cultivability accordingly. But even when advanced techniques are available, the amount of cultivable land may remain very restricted. For instance, of the total land surface of the Republic of South Africa little more than an estimated 5–10 per cent can at present be cultivated. In desert areas like the Saharian zone, the potential for cultivation depends on the availability at low cost of scarce water resources (see p. 205). The percentage of available cultivable land is therefore much less than in South Africa. The highest percentage of cultivated land can be found on the rich soils of East Africa such as the Kigali district of Uganda and the Kikuyu Highlands of Kenya.

Problems of shifting cultivation

'Shifting cultivation', or the bush-fallow system, evolved over the centuries in the tropics, enabling cultivators to obtain a relatively stable level of food production without destroying soil productivity. Using this method the farmer clears small plots of land in forests or savannas on a rotational basis, ultimately leaving the exhausted land to lie fallow for a period during which the natural vegetation and soil fertility can become reestablished.

This primitive system however, is not now capable of meeting the increased agricultural production needs required by rapidly growing populations, for by this method the land produces on the average only 3–5 years out of every 15–25 years, and such crop yields are low by modern standards.

It is very difficult to say where shifting cultivation begins and ends, and the difficulty of definition is increased by the fact that many systems cover a wide range of soils used in different ways. For example cultivation varies considerably between forest and savanna, the rest period being longer in the latter environment. The shifting cultivator usually has an understanding of how his environment is suited to his needs: he can judge the fertility of a piece of land and its suitability for one or other of his crops by the vegetation which covers it and by the physical characteristics of the soil.

From this it can be seen that there are tremendous variations in the types of crops grown, the length of cropping and of fallow periods.

A shifting or slash-burn type of agriculture occurs in most tropical rain forest areas. This system consists of cutting trees and shrubs at the end of the dry season, letting them dry and then burning them. After burning, there is a significant increase in soil nutrients due to the minerals, particularly phosphorus, calcium and potassium, deposited by the ashes. The enriched soil is then prepared for planting corn, beans, cassava and other crops.

However, when the forest is cleared in this way, the natural cycle is disrupted: vegetation is destroyed and plant litter, which decomposes under high temperature and moisture, is not replaced. As a result, released nutrients are rapidly leached from the soil, and soil fertility decreases.

Under circumstances of relatively mild population pressures and availability of adequate land, shifting cultivation has been and still is an efficient means of food production. However, heavy population pressures in areas like West Africa, cause the rotation period to be shortenend by reducing the time in fallow, resulting in an inadequate length of time for restoration of soil fertility and consequently in serious detrimental effects on both land and productivity. One spectacular result of such practices is that the area of closed tropical high forests south of the Sahara has shrunk by at least 100 million hectares. It is obvious that we can no longer be complacent about the continuance of a practise which is so destructive to soils and forests.

Unfortunately little progress has been made towards constructively modifying the system of shifting cultivation.

The elimination of such an ancient practice would require the breaking down of traditions which have been appropriate for centuries, and the problem of providing attractive alternative agricultural systems is far from easy. Although most of the practices of this type of cultivation have become destructive to the land under conditions of increasing human population, the system still has much value in low density communities where the land can be left in fallow for sufficient periods of time. Under such favourable conditions a variety of habitats and valuable 'edge effects' result: the plots which are left fallow are rapidly invaded by various grasses, herbs and shrubs, offering ample food supplies to many species of rodents and other forms of wildlife such as forest antelopes. These are often harvested by the farmers, thus providing him with a secondary protein crop.

The disadvantages of shifting cultivation are perhaps more obvious than the advantages: great expenditures of labour are needed to clear the site; generally low yields are associated with the system; a considerable proportion of the total land is in fallow and therefore not productive at any one time; the system is stable only as long as fallow periods can be maintained for an adequate period; and last, but certainly not least, under the heavy population pressures prevailing today this system has proven to be inadequate and too destructive to the environment.

The advantages of the system under favourable conditions are: the annual cycle of work imposed on the farmer is dictated by and adapted to the seasons; the dry season allows cleared material to dry out and be burnt before the rains; newly planted crops benefit from the ash derived from burning; and clearing is incomplete which does leave some natural vegetation which after desertion of the land can more quickly re-establish itself.

The value of this system should not be dismissed out of hand. It is a highly advanced system from the point of view of empirical ecology and it could be maintained in certain soils where pressures on the land are not excessive.

Subsistence farming

The principal basis of subsistence farming is the production of crops for local consumption or barter, rather than for export. The conception of agriculture as a profitable industry is relatively new to the indigenous Africans. As human populations steadily continue to increase, greater emphasis will fall on the local production of food supplies.

Subsistence farming varies in method and type of crop according to the nature of the country, the possession of stock, and the custom in regard to consumption of food. Although the cultivator is traditionally conservative, he is not so conservative as to cling to methods which experience has shown to be unfruitful, nor does he resist the introduction of new crops.

Africa's principal grain crop is maize which after introduction from America by the Portuguese spread across the continent with amazing rapidity. Although enormous quantities are now grown, very few areas produce enough surplus for export. East Africa is the main production area, but Nigeria and Zambia are also big producers.

Millets of various species were the principal cereal grown in tropical Africa before the introduction of maize. Millets (*Pennisetum* and *Eleusine*) and sorghum (*Sorghum vulgare*) are the basic food staples of the people inhabiting the drier regions, particularly the Sudanian and Sahelian zones, but also East Africa.

Until recently the seeds of fine-seeded annual grasses were eaten widely. This practice now continues in relatively few regions, other grains such as millet having taken their place. In Ethiopia, however, teff (*Eragrostis abyssinica*) is still the staple bread grain.

Rice is increasing in importance as a cultivated crop in tropical and subtropical parts of Africa. Lowland rice grows well in the coastal deltas of West Africa, but also occurs in East Africa. Upland rice is grown in the Sudanian and Sahelian zones on land that can be easily flooded. Roots and tubers such as cassava, Chinese yam, sweet potato and taro are the staples of both the Guinean zone and the low laying areas of East Africa. The cultivation of potatoes is relatively unimportant and restricted mainly to East and South Africa.

Cassava, which grows abundantly in most of tropical Africa, is the crop that is most heavily relied upon in times of drought and famine. Although its nutritional value is low, it produces more calories per ha than any other crop. Experiments have shown that a yearly average of 30 tons per ha can be produced, which is considerably more than is presently the case. Among the indigenous legumes or pulses there are only a few of general importance.

The best known are the cow pea (*Vigna unguiculata*), the pigeon pea (*Cajanus cajan*), the Bambara nut (*Voandzeia subterranea*), and Egyptian clover (*Trifolium alexandrinum*). The groundnut is valued as a foodstuff and its oil is used for domestic purposes. There are numerous subsidiary crops such as pumpkins, cucumbers, wild spinage, tomatoes, peppers, and occasionally sugarcane.

Most of Africa produces a wide range of fruit and vegetables. Market gardens are still relatively uncommon, but more and more attention is being drawn to them. The quality of many fruits, citrus fruits in particular, can be much improved. In general Africans have little interest in growing fruit trees which require continuous care.

African farmers continue to use a considerable variety of edible wild plants, including roots, mushrooms, and fruits. The coconut palm has a wide distribution throughout Africa and the flesh of the coconut and the oil produced from it are widely used.

The influence of colonial powers

Although the influence of colonial powers has diminished in Africa, it seems useful at least to outline some of the major effects which the colonial era wrought. Whatever the political indignities of colonial rule, it cannot be denied that this system has brought certain material assets to many countries.

However, whilst prior to colonization Africans caused relatively little and localized degradation of the land with their primitive methods of agriculture, the picture changed rather abruptly under the influence of the colonial powers around the middle of the nineteenth century. Changes were not only restricted to accelerated land degradation, but included also changes in the spiritual, social and economic aspects of life. Many problems of pastoralists derive from this era when Europeans imposed a new order in which the social organization suitable for nomadism no longer had survival value or when they introduced methods of transport and exchange which disrupted their societies. Although the economic considerations of the colonial powers were aimed mainly at providing additional revenue for their respective governments and expatriate entrepreneurs, local populations did benefit from the increased availability of foreign exchange and circulation of money provided by expatriates.

The agricultural economy altered considerably with the introduction of industrial plantations and the development of new patterns of indigenous agriculture, such as the cultivation of single crops, rather than mixed crops. The former development particularly has often had a detrimental influence on the fertility and therefore the productivity of the land and it seems justified to say that in general the colonial era precipitated and accelerated the process of land deterioration through the initiation of such plantations. In all fairness, it should be recognized that well-managed plantations do exist and there are many such rubber and oil palm plantations which do not cause environmental deterioration and produce considerable external exchange for the countries concerned.

With regard to indigenous agriculture, new plants (see p. 135), improved methods of cultivation, and new methods of plant disease control were introduced. Many of these activities have been very beneficial, but the application of pesticides – a 'gift' of the colonials – can be extremely hazardous if not carefully supervised (see p.158). Attention was also paid to animal husbandry, and the improved methods of disease and parasite control were responsible for much reduced mortality figures of livestock. This, together with the provision of better water supplies and improved breeding steadily increased livestock numbers. However, such increased numbers have resulted in increasing pressure on the range and its subsequent deterioration (see p. 123).

There are many countries in Africa where European settlement was either unattractive or actively discouraged.

In such countries, even until the outbreak of the Second World War, active planned development was rarely regarded as a desirable thing by colonial governments.

Local agricultural products were exported and sold by expatriate entrepreneurs and sold in exchange for imported manufactured products (CLAWSON, 1965).

The dual system of colonial paternalism and laissez-faire private enterprise had effects which put an end to the Garden of Eden picture of preserving everything as it was, and has now led in many parts of Africa to rapid deterioration in the use of land, and to some extent in the whole structure of society (CLAWSON, *op. cit.*).

The effect on the land of European immigration was such that population pressures came more prominently to the forefront. An undeniable benefit to Africans has been the discovery by European pioneers of the best way to use land under African conditions, the introduction and breeding of improved stock, and the systematic production of crops like coffee, tobacco, with a good world market value (CLAWSON, *op. cit.*).

Cash crop production

Production of cash crops on a large scale began in the early part of this century when land was still abundant. Hundreds of thousands of acres of forest and fallow land were cleared primarily to produce such crops for export. As a result, both land and labour available for domestic crop production declined.

In response to world trade demands many national agricultural programmes now concentrate on cash crop production and the rapid pace of development in this field is quite unprecedented for Africa. The volume of produce that emerges from Africa is very considerable, although the variety of principal export crops is limited and includes mainly oil palm products, cocoa, cotton, coffee, tea and groundnuts. Synthetic products are now competing with cotton, sisal and natural rubber. The production of sisal has been seriously affected by this. The first important agricultural export from Africa was palm oil, exported from Nigeria in small quantities as early as the eighteenth century. Cotton, coffee and groundnuts were known in Africa long before Europeans had penetrated beyond the coastal fringes, but they became important exports only in the early part of this century.

Africa's agricultural exports today constitute a sizeable contribution to world supplies. Oil from palms, for example, represents over 70 per cent of world production (FAO, 1968), and although it is still produced from the fruits of trees growing naturally in forests, by far the greatest volume now comes from higher yielding plantation stock. Coffee is both a plantation and a peasant crop, most of the coffee being produced on small plantations of indigenous peoples. The *robusta* variety is widely grown because it is disease resistant, but the *arabica* variety, which is believed to have originated in Ethiopia, is more subject to diseases and is only found locally in East Africa. Africa also produces most of the world's cocoa (805.000 tons in 1967/68, FAO, 1968), as the vast areas covered by humid tropical forests of West Africa provide excellent growing conditions. Most of the cocoa is grown by small landholders, and the productivity level is relatively low due to poor management, parasites, and diseases.

Tea is grown mainly in East Africa where the biggest producing countries are Kenya and Malawi. Until recently this was a plantation crop, but is grown now with increasing frequency by farmers. African tea represented 12 per cent of world production in 1968 (FAO, 1968).

Natural rubber production is restricted to the West African rainbelt and contributes only seven per cent of total world production (FAO, 1968). Liberia is the only country where extensive plantations exist; smaller plantations are found in Zaïre and the Ivory Coast. Although there is some competition with synthetic rubber, it is expected that these plantations will remain productive. Two main varieties of cotton are produced in Africa, the fine quality 'long staple' in the north, found particularly in Egypt and the Sudan. and the 'medium staple' found elsewhere, particularly West Africa. Production is increasing, notably in Nigeria. Leaf tobacco is also grown in many parts of Africa, Rhodesia being the principal producer and Malawi running second. Pyrethrum production is generally restricted to East Africa. World demands for this product have diminished in recent years.

The groundnut is of major importance as an African cashcrop and production is centered mainly in the countries of the West Coast, where West Africa is the largest peanut exporter in the world. Cane sugar is grown in many parts of Africa but is cultivated on a large scale only in a few countries in Eastern and Southern Africa. Production is not high, and importation of sugar is necessary. Bananas are grown very widely throughout Africa but export contributes only 5,5 per cent of the world total export (FAO, 1968). However, they are an important export crop for the Ivory Coast. Both plantains (*Musa paradisiaca*), used for cooking, and the familiar table fruit (*Musa sapientum*) are grown. Highly weathered tropical soils are often not as fertile as the high natural forest suggests and the use of tree crops in the forest areas should therefore be emphasised. As compared with most annuals and biannuals they make relatively small demands on the soil nutrients and avoid losses due to successive stages of clearing, burning and cultivation. Once established, they protect the soil from the destructive effects of the climate and anchor it with their roots much as did the original forest vegetation; they also feed it with their litter. In a well-established rubber, oil palm, cocoa or coffee farm one can visualize a fairly closed cycle of nutrient circulation, even if the amounts circulating are lower than under the original forest. When tree crops become too old, they are relatively easily replaced by the same or another crop which is introduced beneath the former in order to minimize soil exposure. Tree crops thus protect and benefit the soil, while the farmer in turn benefits from the fact that once they are established, routine weeding and maintenance is small compared with the laborious job of clearing new land each year. He also has a permanent capital asset for his labours.

There is a basic conflict between the production of cash crops for export and production for local consumption. As more land is used for cash crop production, less is available for food production. This conflict is particularly acute in countries where fertile soils are favoured for cotton and groundnut crops. With rapidly increasing populations these soils are also badly needed for the production of human food. On the other hand, foreign exchange from cash crops is badly needed by most African governments for their national development.

Although cash crop production generally depletes the fertility of the land, there are fortunately exceptions. The rubber plantations in Liberia are an excellent example of how under good management both the production of rubber and the fertility of the soil can be maintained.

The Ground-nut scheme failure in Tanganyika

Failures in land use have often been spectacular in various countries of Africa but perhaps none has been more widely criticised than the ground-nut scheme of the British Government. This was devised by an executive of the United Africa Company. On the basis of a series of brief surveys that began in 1946, the U.K. Ministry of Food and Agriculture advanced a programme for transforming about 12,150 km^2 of Tanganyika – then a British territory – into a ground-nut cultivation area for the production of margarine and oil destined for British consumers.

The plan turned out to be a costly failure. The project was regarded as mainly an engineering exercise and an inadequate amount of field work was done to assess the environment – suitability of soils, natural vegetation, local micro-climate, availability of ground-water and so on. Virtually no ecological study preceded the selection of sites and the soils, although productive, dried out so hard in the sun after cultivation that the crops were imprisoned in them and gully erosion occurred in the tracks of the tractors. Administrators of the programme failed to realize the great variability of African soils unaltered by agricultural practices and disregard of this factor was one of their major downfalls. This fact was not brought home until costly experience made it apparent that the soils naturally suitable for cropping were often limited in distribution and widely dispersed. Bush clearing was done in blocks of 2,59 km^2, without regard to soil variation, and this resulted in the clearing at high cost of very large areas which later proved to be quite unsuitable for annual cropping. The area actually cultivable turned out to be small, not only on account of soil composition but also because of erratic rainfall, too uncertain to allow for cultivation with safety. Soon after the scheme was recognised as a failure it was dropped and the production of ground-nuts by large-scale mechanized farming methods has never been revived in East Africa (RUSSELL, 1972).

Despite this glaring example, fallacious assumptions regarding cultivability continue to be made and too often inadequate attention is given by preliminary survey missions to the ecological resource base of the land. Where mechanization is premature and unlikely to improve much on local practices of production, it should not be encouraged beyond a desirable minimum. To decide when and where to introduce this form of farming demands an appreciation of the ecological setting, promises and hazards of the locality and an understanding of the local people (PHILLIPS, 1959).

The failures described here are only illustrative of similar developments elsewhere on the Continent. Another consideration is that the objectives of agricultural development projects, often heavily subsidized by multilateral and bilateral aid, ultimately prove unsuccessful from the point of view of increasing per capita food consumption, simply because there are too many hungry mouths to feed. In such cases there seems to be little purpose served in the exercise.

What are rangelands?

Rangeland may be defined as any land which is unsuitable for permanent arable agriculture or productive forestry and which produces native forage for animal consumption. As defined, it includes not only the drier savannas and grasslands receiving less than 700 mm of annual rainfall, in which cropping is either unreliable or impossible without irrigation, but also mountainous or hilly topography and many swampy or badly drained areas in those regions receiving more than 700 mm annual rainfall. On this basis, if one includes the deserts, Africa consists of 8 million square miles of rangeland of which 5 million is grassland or savanna. However, not all authors are in agreement with this definition and include more productive rangelands potentially suitable for cultivated lands and forests. In most rangelands, a combination of drought and overgrazing leads to deterioration of the soil and erosion becomes pronounced leading to further degradation of the vegetation and sometimes even to the development of subdesert conditions. Thus subdesert or even desert vegetation is found in places where climatic conditions are in fact quite suitable for more highly evolved communities.

Effects of overgrazing on rangelands

The extension of uncontrolled grazing and animal husbandry into arid and semi-arid zones often has a most harmful effect on the environment, creating a marked ecological imbalance. Unfortunately this malpractice occurs over most of the grazing lands of Africa. Livestock being selective eaters, choose the more palatable and nutritious grasses, and unless proper stocking is practiced, these are gradually replaced by less desirable plants. Once competitive vigour of grasses is reduced by overgrazing, the establishment and growth of trees, shrubs and weeds follow, in turn reducing the production and availability of grasses and lowering the carrying capacity of the range. The accelerating expansion of woody, and often thorny growth of little value to stock is thus the curse of many grazing areas of Africa.

Chronic overgrazing does inevitably modify the original herbacious cover. Perennial grasses are killed and are replaced by plants whose nutrient requirements are lower, or which successfully evade the effects of overgrazing by various protective devices such as thorns or unpalatable bark. This results in a smaller turnover of organic matter. As grazing continues, the less valuable species will bear the brunt of the grazing pressure and their numbers will decrease. These plants are not followed by new invaders, but, rather, the land approaches a barren state, with soil regressing rapidly.

If excessive grazing continues, the population levels of even these species will ultimately be depressed. With further degradation of the environment the carrying capacity for the various species will be reduced and a lower biomass production will result.

In general overgrazing also tends to modify the composition of the insect fauna of arid regions by precipitating the disappearance of mesophyle species and the multiplication of xerophyle species. The increasing problem of locust and cricket damage is caused largely by overgrazing. These insects can complete their life cycle better under such conditions and may develop to plague proportions (see also p. 151).

During the late nineteenth and early twentieth centuries vast changes occurred to the vegetative cover throughout the African continent as a result of overgrazing. In South Africa, natural grass, tree and woodland savannas have been largely replaced by trees, shrubs and herbs typical of a more arid climatic regime and these in turn have accelerated erosion.

In Karamoja, Uganda, misuse of the land has caused serious disruption to the ecosystem and much of the dry thicket existing there now derives from savanna or steppe as a result of overgrazing. The following shows in diagrammatic form the various processes that have taken place (after LANGDALE-BROWN, 1964):

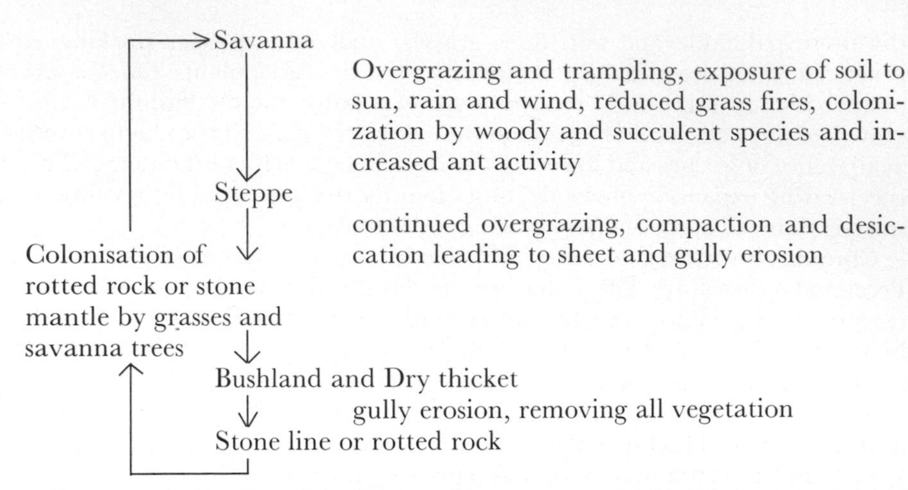

It appears therefore that the savannas, steppes, bushlands and thickets of Karamoja may be stages of a cyclical succession induced by overgrazing!

Effects of degradation of the environment on the productivity of wild herbivorous animals

When the influences of cultivation, fire, and grazing become over-accentuated, deterioration of the grassland communities sets in, resulting in lower and less advanced types of biotic communities and, eventually, in serious destruction of the habitat.

An indication of the serious extent of land decline can be found reflected in the continuously decreasing wildlife populations. While an overall decrease in wildlife is undoubtedly taking place, some species do temporarily increase because the plant composition of the degraded rangelands is more suitable to their needs. Impala, springbok and oribi are examples of this situation (see p. 95).

An interesting phenomenon which has been observed is that in overgrazed conditions high populations of certain rodents and lagomorphs, such as hares, occur. Contrary to what was earlier considered to be the case, that these so called pests were a cause of range depletion, many recent studies have indicated that instead their high numbers are themselves a consequence of range depletion by livestock grazing.

Land use and soil erosion

The process of soil erosion in Africa has been well documented by HARROY (1949) and a map showing the varying degrees to which the danger of erosion exists throughout the continent has been compiled by FOURNIER & D'HOORE (1962).

The term 'erosion' as it is generally understood has too limited a connotation to describe a process which results in radical changes in the whole character of the land; loss of mineral plant foods, oxidation and disappearance of organic matter, breakdown of soil structure, degeneration of vegetation; and the setting up of a new train of land and water relationships. The whole complex process of destruction is best referred to as land degradation (ALLAN, 1949).

After more than half a century of agricultural efforts by Africans under the guidance of expatriates, standards of land use in many developing nations unhappily show a general decline, and soil fertility continues to decrease. An advanced state of soil degradation, with increasing soil losses by erosion has already been reached in the overcrowded regions where the traditional systems have collapsed and have been replaced by continuous cultivation.

Throughout Africa, when the rural human population and the associated livestock population has passed a certain critical density, the soil is destroyed with devastating ruthlessness.

The African continent now is faced with the worst erosion problem in the world. In Europe, which has the least erosion problem, the loss of earth equals on the average 84 tons km^2 per year, whilst in Africa the figure is 715 tons km^2 per year (FOURNIER, 1972).

Serious erosion is generally associated with careless clearing of the land and bad farm and road layouts. Unfortunately this has all too often been the case in many parts of Africa.

Erosion proceeds most rapidly in places where the topography is undulating and with slopes of more than 5° rapid truncation of profile may occur because of soil slippage. Continued erosion may lead eventually to the complete removal of the soil and the exposure of a resident stone mantle. It does not usually continue into the underlying rock, but ceases when the stone line, or rotted rock layer is reached.

Two areas in Africa are particularly vulnerable to erosion. One area is the tropical belt in which most soils are very fragile. Here, any injudicious clearing of the vegetation exposes the soil to torrential downfalls of rain which leach out nutrients. When the rains terminate, either wind blows away the topsoil, or the sun bakes the soil to such an extent that plants have difficulty pushing through the surface crust. The other area susceptible to erosion is the arid belt. Here temperatures of over 38°C are common and daily temperature fluctuations are also extreme. The soils lack humus and have a poor texture. Evaporation of the already scarce water, which leaves only salts behind, is a constant hazard.

It is clear, therefore, that in vast areas of Africa the soils suffer from many more environmental risks than those of the temperate lands. Fortunately the continent is also blessed with some very productive soils, but it is on these that heavy concentrations of people unfailingly occur. Under good management these soils should be able to support these concentrations, but unfortunately management is rarely optimal and the good soils are therefore also in some danger of losing their fertility.

Crop failures are a very common feature in Africa due to unreliable rainfall and lack of soil fertility, resulting not only in mere financial losses to the cultivators but also sometimes in starvation and the destruction or dispersal of the communities concerned. The implications are clear: if the hoe cultivators of the weak soils that cover much of Africa do not acquire a better understanding of how to cultivate and conserve their soils, their future will be in serious jeopardy.

The most serious consequences of erosion are impoverishment and biological deterioration of the soil, for erosion attacks the biodynamically effective layer of the lithosphere. It is at this level that the last stages of the recycling of a number of mineral elements take place and where nutritive elements accumulate which are basic factors of fertility. The disappearance of this layer of soil, often linked with poor protection by vegetation, obviously has an effect on the productivity of ecosystems, and with this top soil loss there is further degradation through the creation of washes and gullies.

Erosion is responsible for removing the finer parts, clayey and humid colloids from the soil, thereby affecting the absorbing complex and, in consequence, the soil's content of fertilizing elements. The result is a drop in agricultural production which can only be corrected by an increased use of fertilizers and other improvements, but usually these corrections are not completely effective and, moreover they are expensive in their application.

One typical example of the consequences of erosion was given by FOURNIER (1963) for Niangoloko in Upper Volta, where he calculated that a change in the annual erosion rate of 143 tons km^2 to 1318 tons km^2 resulted in an annual loss of clay and humus passing from 26,5 tons km^2 to 49,5 tons km^2. The millet yield dropped correspondingly from 729 kg per hectare to 352 kg per hectare. At the same time, the annual run-off was tripled, which fact brings to mind another consequence of erosion – that water does not effectively percolate because in eroded regions the permeability generally decreases and the areas have therefore an even greater need of water.

Damage from erosion in southern Africa is most severe in Malawi, Swaziland, Leshoto and Botswana. In North Africa it is severe in all countries. It is also serious in most of East Africa, and particularly in heavily populated areas such as Ruanda and Burundi. In Tanzania, Kenya, Uganda and Somalia there is the very serious problem of erosion by water, combined with water shortages during the dry season, which aggravate wind erosion. Thus, the land must suffer additional depletion if nothing is done to conserve the soil and water. In the northern part of Somalia, for example, there are no rivers, but instead there are wadis or tugs – great natural drainage channels which gash and furrow the face of the countryside. For most of the year they are dry, but when the rains fall, they become rushing brown torrents, down which 3 or 4 million tons of water may pour in a day. Like the springs, they seep away into the parched earth. Thus a terrible type of water erosion exists in a country so depleted of vegetation as to be now mainly a semi-barren land.

Soil erosion, whatever its cause, gradually makes the land unhabitable.

As the soil becomes depleted by erosion, people naturally attempt to move to other, more productive land. Eventually, when there is no more land available, they are forced to adapt themselves to smaller amounts of food which require harder work to grow. This condition eventually leads to malnutrition and hopelessness. It is a situation invariably associated with severely eroded land where large numbers of people must eke out an existence.

In the semi-arid areas of North Africa wind erosion hazard causes the greatest problem. Here, rainfall is generally of high intensity and extremely seasonal in nature. The rainy season is followed by a period of many months when there may be no precipitation whatsoever. When a reservoir is established in such lands a fairly predictable sequence of events takes place. The increased water supply, usually without control of stock numbers leads to an increase in the number of cattle, then overgrazing and erosion (see also p.123). Subsequently, a great deal of the sediment carried by the flash-flows is deposited on the bottom of the existing reservoirs which gradually become silted up and in time useless (LE HOUÉROU, 1970).

Often the pattern of stock-routes changes when a new reservoir is constructed. Large areas around it start to suffer from increasing pressure on soil and vegetation by the vast herds of cattle which are directed towards the new source of permanent water. The ground cover is thus soon removed and erosion sets in.

It is obvious from the foregoing that with the accompanying degradation of the surrounding land and the eventual silting up of the reservoir more damage than good is wrought!

IV. SPECIFIC ENVIRONMENTAL PROBLEMS

Nomadism and consequences of sedentarization

Nomadism is an adaptive response to an inhospitable arid environment. It is pastoralism on the extensive scale, following available forage and water. All too often nomads are dismissed merely as wanderers, which they certainly are not; rather, they are close observers of seasonal home ranges (DARLING & FARVAR, 1972), and in fact, the word 'transhumance' refers to the definite movements of nomads living south of the Sahara from their 'winter homes' in the north to 'summer homes' in the south. These movements do not generally exceed 200 miles.

Nomadism is essentially a type of subsistence land use that is becoming more and more marginal to the market economy. Thus, many land use economists believe that the system should be discontinued. However, ecologists and specialists in animal husbandry believe that this system has been and could continue to be an effective means of harvesting a highly variable forage supply.

While the features of nomadism and transhumance are admirably suited to nature's changes, they present by their very nature a built in resistance to development and orderly administration of the rangelands into an acceptable market economy. The nomads have also certain prejudices or customs that are not conducive to the development of such an economy. For example, if the animals serve mainly for prestige purposes, such as purchase of brides, as they so often do, or merely meet immediate food needs, there is little incentive to secure quick growth of animals for early marketing or to increase the off-take.

The system of nomadism in particular can be easily upset or broken by increasing population pressures of both man and beast, and as it is utterly bound up with movements of man and his stock, the habitat cannot be adequately conserved if large numbers are involved. Excessive lingering punishes the vegetation, reduces the number of species of plants present on the range and of course impoverishes it.

BROWN (1971) states that there are probably 50 million pastoralists in Africa south of the Sahara. These people depend for their subsistence wholly or largely on the products of their stock: milk, meat, blood and hides. Brown points out that in semi-arid areas where pastoralism is the most suitable form of existence, pastoral people must, of necessity, keep sufficient numbers of domestic animals for their own subsistence which often exceed the level the environment can support. If a population of domestic animals is high in relation to the carrying capacity of the environment at the start of a wet period, the numbers will increase rapidly during this period and when a drought year follows, the population grossly exceeds the capacity of the range which is reduced under the adverse circumstances. Unless therefore there is a deliberate take off of livestock from the affected range during the drought, not only can the population be expected to suffer heavy losses through starvation, but at the same time the range suffers serious damage.

Drought may kill off 40 per cent or more of local herds in seasons of exceptional severity and in regions of poor and uncertain rainfall periodic drought imposes a check on livestock numbers as harshly and dramatically as any epidemic disease.

BROWN (*op. cit.*) states that when we speak of overgrazing by domestic stock, we are often in fact observing a situation in which the human population density is excessive in relation to the carrying capacity of the environment. To live, pastoralists are forced to keep sufficient stock for their subsistence, and in doing so may exceed the carrying capacity of the environment in terms of the primary user, the stock – thus causing widespread destruction. Some of the classic cases of 'overgrazing' are due to overpopulation by humans following the establishment of law and order by colonial administrations, and the control of disease and famine.

LIVESTOCK CARRYING CAPACITY AND LAND REQUIREMENTS OF PASTORALISTS

The land requirements of a pastoral community are determined by two factors: the number of livestock needed for subsistence of a family or an individual, and the stock carrying capacity of the grazing land. The latter factor is enormously variable, and depends on the average acreage of a pasture land which is capable of maintaining an animal for an indefinite period without deterioration of the pasture. This, in turn, depends mainly on soil and climate, but also on the system of management.

Movements of nomadic flocks and herds are determined by various factors of which seasonal distribution of water and grazing are most important, while other less obvious factors such as tick-borne and fly-borne diseases, biting flies which make life almost intolerable for cattle, and the occurrence of natural salt, modify the pattern and are of major importance in some areas (ALLAN, 1965).

During the dry season the stock must remain within range of the limited permanent water supplies, but with the start of the rains the bulk of the flocks and herds move out. However, not all the livestock move with the rains. Considerable numbers remain near the permanent water sources throughout the year with the result that these areas are never rested and consequently deteriorate rapidly. The size of the basic herd will be determined by the type of animal, by the way in which the herd is exploited for food and by the environment, to the extent that environmental factors affect the productivity of each animal. The number of livestock required to maintain a wholly pastoral existence varies considerably. Thus the Fulani live almost entirely on milk and its products and as they keep no goats or sheep, they must maintain a very large number of cattle per human unit. Other nomadic tribes that keep mixed flocks of cattle, sheep and goats require far less cattle beasts per capita. Once the road to agriculture has been taken by the pastoralists there is, generally, no return, for the practice of cultivation allows a population increase well beyond the limit that simple pastoralism can support. Consequently, wherever conditions are favourable for crop-production, agriculture tends to displace pastoralism, either by the conversion of the pastoralists to cultivators or by the encroachment of cultivating peoples on their grazing lands (Allan, 1965). Thus, over the centuries, community after community of Nilotes, Nilo-Hamites and Fulani have taken to the hoe and exchanged the freedom and austerity of nomadism for the greater security of a sedentary or semi-sedentary way of life. Pastoralism, in the sense of complete dependence on flocks and herds, survives mainly in regions marginal or unsuitable to agriculture, where the conditions that discourage crop production also impose the necessity of nomadism on the pastoralist.

Ecological consequences of sedentarization

Serious ecological consequences arise from the sedentarization of nomads. Wherever these people are forced to settle, they acquire the habit of lingering on with their stock around wells or boreholes, causing overgrazing and resulting in a decline in the quality of livestock. Even where the stock does not exceed land capacity, other affects of sedentarization become apparent in the form of poorer conformation and a rise in deficiency diseases and parasitic infections (Darling & Farvar, 1972).

Some governments, like the Sudanese, now realize that much of their land area would be uninhabitable if it were not under nomadic pastoralism, being unsuitable to most forms of agriculture. As many mistakes have been made in development and settlement schemes, a re-appraisal of the fate of the nomads and an ecological assessment of the land occupied by them is much needed.

Perhaps new solutions can be found by which a nomadic existence can be maintained with, however, greater emphasis placed on health, welfare and education. It is, for instance, possible to improve the transhumance system in such a way that pastures are improved on both the wintering and summer grounds, which combined with improvements in social services would considerably improve the standard of living of the people involved.

Marginal lands

The insidious process by which productive land turns into unproductive marginal land has much accelerated in recent decades. The term 'marginal' is applied to any area which, for reasons of climate, soil, elevation, accessibility or distance from markets, is for the undertaking of major and costly measures of amelioration of doubtful economic value (Whyte, 1966). It seems that the trend over the past hundred years has been for more and more land to move from the productive to the marginal category, and from marginal to completely unproductive land. It is a trend which is especially noticeable in semi-arid and mountainous countries. Growing pressure on land resources has led to the extension of crop farming into areas which are not normally suitable for such production.

Much marginal land is now in that very condition due either to past mismanagement, or to the application of ecologically unsuitable forms of land use which has downgraded land to the point where it is unsuited to or marginal to present forms of use. It can no longer be used for agriculture, forestry production or grazing under prevailing economic and technological conditions. The significance of mountains stripped of trees and soil until only rock remains, or of the recently spreading dunes of the Sahara is that within the foreseeable future, that is, within the next few decades or even centuries, this land will not recover its original state of productivity. It is particularly this tendency towards irreversible loss that provides the greatest motivation to agencies and governments concerned with land improvements to halt this unhappy trend.

With the human population ever on the increase, we are faced with an urgent need to halt such downgrading trends and to bring the land back to some acceptable level of stability with the ultimate aim of producing permanent levels of productivity. This, if at all feasible, is a time-consuming process which will require not only considerable international assistance to the Governments concerned, but also co-operation of the local people and a rapid change in their attitudes towards land use. If these people cannot come to appreciate the reasons why efforts towards land improvement are made, little will be accomplished. It is very necessary therefore to educate the people along these lines. It will also be necessary to assist them financially during the period when the land improvement practices are set in motion.

Probably the world's largest, continuous stretch of marginal land is the southern edge of the Sahara desert. It is not a clear edge, but rather a zone fluctuating in size between a few kilometers and some 500 km deep, stretching 5000 km from the Atlantic Ocean to the Red Sea. Inside this zone, and characteristic of all marginal land in semi-arid areas, are bare patches of soil that grow larger yearly; sand dunes which are forming and spreading, and streams which are drying up as the border of the Sahara moves southward, turning even more land into desert.

Although, in dealing with the marginal land situation in Africa, priority should generally be given to arid regions, certain mountainous areas should also be included in any future programme for stabilization and reclamation, particularly the arable slopes of mountains within the semi-arid zones and the hilly and mountainous areas in the humid tropics. These mountainous zones if downgraded to marginal conditions are not only themselves affected but in turn affect the plains at their feet by deposition of erosion material.

The most serious marginal lands problem in Africa is that of the arid and semi-arid lands. It is here that the lack of sufficient water in addition to a high salinity level in the soil are the most overriding restraints to man's use of land.

Of the potentially limiting factors-topography, soil and climate – in an arid environment, the most important is climate or rather limited rainfall. It is therefore imperative that land use practices are geared to this. In East Africa, for example, rainfall is seasonal and erratic; if there is a rainfall expectation of over 750 mm per year, there is a reasonable prospect of harvesting crops. However, if the expected rainfall is less than 500 mm per year, crops should not be grown without irrigation and the ranching of cattle on an extensive basis is the only possible form of economic land use under dry-land conditions.

In the lands of uncertain or markedly seasonal rainfall, the tenure of man has always been insecure. Climatic fluctuations in the arid zones have both short and long-term effects: in the short term they incite the cultivators during the more humid periods to extend their activities and the pastoralists to transgress the boundaries of the areas which they would normally occupy during dry years; in the long-term these expanded activities of the pastoralists and cultivators cause increased environmental degradation because during dry years the activities cannot be sustained and result in increased erosion and overgrazing with the consequence that after a period of weather stress the land cannot readily recover its former ability to grow useful plant cover and to absorb moisture.

Due also to this marked variability of rainfall, villages and settlements have sometimes simply disintegrated as can be seen from the many ruins of settled communities dotted across the arid lands.

Within the dry lands human occupation has, over a long period, broken down from time to time and there has been a progressive shrinkage of habitable land because of salination.

With few exceptions – such as some arid areas in South Africa – little development has taken place so far which takes full advantage of modern technology and it is most essential that further salination which results from bad irrigation practices, be stopped.

In the past, within the dry areas populated by pastoralists accustomed to primitive forms of cattle management, all too often there have been too few watering points, and severe overgrazing has ensued in the vicinity of the water source. However, even in the well planned schemes, it has been found impossible to control the number of stock involved with the same end result ensuing. Furthermore, overgrazing in the catchment area of a dam can so increase the amount of run off that the consequent surge of water in heavy storms may seriously imperil the water reservoirs and rapidly reduce their storage capacity through the increased rate of siltation.

In the very worst circumstances, when good engineering advice has not been obtained, all that is left of a water supply project is a breached dam set in a severely overgrazed area. Unfortunately quite a few examples of such destroyed structures can be seen by any visitor to arid range-lands. The lessons to be learned have been emphasized again and again: the provision of stock watering points *must* be accompanied by a sound grazing plan with some effective means of control over stock numbers and grazing pattern, if not the whole effort is wasted as a great deal of damage is otherwise done to the vegetation and the overgrazed area creeps ever outward from the watering point!

The problem of desertification

The biological productivity of desert environments of benefit to man is expressed in the production of plants for consumption by him and his stock and in fuel for construction and other purposes. The productive potential of deserts has to be considered in relation to the effects on growth and reproduction of those special features of the environment which characterize deserts (see p. 24) and with regard to the ways in which plants and animals are adapted to meet these special conditions.

From an ecological point of view deserts are most interesting environments because of the adaptations which plants and animals possess to deal with extremes in environmental conditions. Desert ecosystems have been inadequately studied and it is to be hoped that increasing emphasis will be given to such studies.

Apart from the Sahara desert there are in the south the Namib and Kalahari deserts, and in the east the Nubian, Danakil and Somali deserts. Man-made subdeserts occur in Kenya, Tanzania, Botswana and Lesotho.

The now widely held view is that the desert areas of Africa are expanding at a disquieting rate. The Sahara is said to be advancing on a 3,000 km front along both its northern and southern boundaries. In the Sahel zone there are areas where, within the memory of the older inhabitants, grass has disappeared and desert succulents have become conspicuous.

Contrasting views have been expressed about the causes of desertification or 'desert creep'. According to some workers this process is the result of a gradual reduction of rainfall since the time when the Romans used North Africa to supplement their food supplies. Ecological evidence from the Saharian zone, such as remnants of a former vegetative cover which survive as post-climax relics, indicates that at least eastern Africa is in a post-pluvial or relatively dry phase which may not yet have reached its peak of aridity. Others believe this expansion of African deserts was precipitated by the destruction of vegetation by man and his livestock. Although there is no indication that precipitation has decreased significantly in recent centuries, yet 'desert creep' continues at an accelerating rate. This seems to be additional evidence that during this recent period man's abuse of the land has been the precipitating factur. It is not certain to what extent the actions of man or natural causes – decreasing rainfall and drift of sand by the Harmattan –

have contributed to the outward creep of the desert, but there is little doubt that destruction of vegetation by pastoralism has increased the effects of natural changes, and that the African deserts are to some extent man-made. STEBBING (1937, 1953) was the first to bring attention to the problem of the expanding Sahara desert. AUBRÉVILLE (1949), and others have also contributed to an understanding of degradation of the vegetative cover in semi-arid zones, particularly the Sahara. Specifically, they found that the impoverishment of the vegetation results in an impoverishment of the fauna, particularly mammals, whose life is dependent upon the existence of vegetative cover and water. LE HOUÉROU (1968) described the progress of desertification in the area north of the Sahara from Libya to Morocco, where the following sequence occurred: a reduction of the perennial vegetation, impoverishment of the flora, soil erosion, the development of moving dunes and the establishment of a desert-like substratum. Le Houérou believed the process was iniated by overpasturing, cultivation of cereals which resulted in the destruction of the natural vegetation, and other destructive land use practices. Several other authors have studied the expansion of the desert into the steppes south of the Sahara. Here fires and the destruction of trees and shrubs appear to be largely responsible for desert creep (see p. 49).

The spread of African deserts is only part of a more general process of encroachment: the desert encroaches on the steppe, the steppe on the savanna and the savanna on the forest. GHABBOUR (1972), for example, describes this for the Sudan: 'A fair estimate of this desertification process can be obtained from the fact that dense *Acacia* scrub, which could be seen in the Khartoum-Omdurman area in 1955, is not to be found at present within 90 kilometers south of Khartoum. Concomitant with desertification is the shifting southward of the gum arabic belt... Signs of progressive desertification can also be observed even in the humid Equatorial Province. During a flight between Wan and Juba, one can see rows of trees growing in the mid-lines of several dry river beds, indicating lower river discharges than what was normal perhaps only half a century ago. Likewise, the lands bordering the Sobat River look so parched from the air that one cannot imagine how they supported 'impenetrable' forests in the 19th century.'

Control of 'desert creep' is likely to be one of the most difficult resource management problems to be solved in Africa and elsewhere because it requires drastic modifications in the land use patterns and attitudes of the relatively primitive nomadic people who inhabit these areas. Control of desertification therefore requires considerable investments in new land-use techniques and education of the people in the application of these techniques.

Reduction in the use of vegetation for fuel will be essential. Fortunately this is possible in some Sahelian countries where readily available petrol supplies can provide a cheap source of fuel. To achieve any measure of success in controlling 'desert creep' the following needs to be implemented in the near future:

1. Drastic limitation of the stocking rate of livestock;

2. Improved availability and use of water supplies and pasture rotation;
3. Establishment of forage resources principally through the aid of irrigation;
4. Control of erosion and restoration of vegetation.

The establishment of stable vegetative cover along the desert fringes should consist of selected, drought resistant trees, shrubs and other plants, which are also good fodder producers. If such measures are not taken it is certain that 'desert creep' will continue and that the inhabitants of affected areas will be severely handicapped, perhaps even to the extent that they can no longer find a suitable livelihood for themselves (see p. 46).

Because of the poverty which prevails in the Sahelian zone a considerable amount of international assistance will be required to stop the process of desertification. The picture is altogether different in Libya and Algeria where some of the oil revenue could be made available for this purpose. However, so far, much too little has been done in terms of international aid.

The invasion of Africa by plants and animals

Man's influence on changing the distribution of plants is simply enormous. The ecological effects of introduced species are very complex, and our understanding of the problem is limited by our scanty knowledge of how even simple ecosystems function. C. S. Elton (1958) clearly described the problem as follows: 'When contemplating the invasion of continents and islands and seas by plants and animals, and their microscopic parasites, one's impression is of dislocation, unexpected consequences, and increase in the complexity of ecosystems already difficult enough to understand let alone control, and the piling up of new human difficulties. These difficulties have mounted especially in the last 150 years, and they have had to be met by means of a series of fairly hasty and temporary measures of relief that are only here and there supported by fundamental research on populations, or even a systematic record of events'.

Man's introduction of foreign plants and animals into the African environment has produced a much greater impact on natural ecosystems and human existence than is generally realized. We have inadequate information about the number of introductions that have taken place in Africa but in the case of plants they may amount to tens of thousands. Among these species only a few have left any major impact on the environment but in the aggregate their impact has been considerable.

Introduced Plants

One of the primary reasons for the introduction of these species has been for crops, gardens or other purposes serving man. The majority of plants now utilized for food in Africa have been introduced from other parts of the world. Some of these, such as bananas, wheat, sugar cane and rice, arrived centuries ago from the East on Arabian ships.

Many plants were brought from the Americas, including maize, potatoes, sweet potatoes, yams, squash, many types of beans, maniocs, peanuts, cashews, pineapples, cocoa, avocados, tomatoes, peppers and papaya. These exotics have been most beneficial in augmenting the food production of Africa, and it is interesting to contemplate how Africans would have fared without them. Certainly their food habits would have been much less varied! Plant introduction is the easiest and most efficient method of improvement of crop production and new varieties of certain plants still are being introduced in Africa to increase and diversify crop production. The possibilities of crop and grassland improvement through such introductions have by no means been exhausted, and there is an increasing need for introductions of new plant material to meet the requirements of an increasing world population. Vast areas of Africa are now covered by plant communities that are very different from the original cover and in most areas exotic plants are the important component.

Yet perhaps the most important general effect of plant introduction is that of progressive impoverishment of the native flora with subsequent decrease of ecosystem stability. Usually the ecological niche occupied by a successful exotic is one from which one or more native species has been partially or completely displaced. One of the most difficult problems in determining this and other ecological effects of introductions is to determine whether an exotic caused a certain effect or was merely able to invade a new habitat due to alteration of the environment by human influences such as fire or intensive grazing. The fate of newly introduced species varies from complete failure – disappearance – to their becoming pests and 'taking over' their new locale. This success or failure depends on the degree of adaptability to the new conditions as determined by genetic and environmental factors.

According to POLUNIN (1960) the adaptations of plants to particular habitat conditions are sometimes so precise as to remain unnoticed, yet sufficient to prevent the leading of an independent life. Often a mere slight change in environmental conditions will threaten the very existence of a species. Unfortunately the ecological impact of pest plants can be substantial: The water hyacinth (*Eichhornia crassipes*) is a free floating native of South America. It has invaded many African rivers and lakes where it chokes out the native vegetation, impedes navigation, and interrupts fish spawning and feeding, causing a sizeable reduction in the catch of fish. It also clogs irrigation pumps and hydroelectric schemes. Another aquatic plant introduced from South America, the water fern (*Salvinia auriculata*) has caused near disaster conditions on Lake Kariba by covering large portions of this huge lake with floating mats.

The North American succulent *Opuntia* and *Lantana salviaefolia* from the Far East have invaded land that has either been allowed to remain idle after cultivation or that has been overstocked for many years. Local problems in South Africa are caused by woody acacias and *Hakea* from Australia which have spread widely, choking out indigenous vegetation.

The ability of exotics to replace indigenous vegetation, vastly altering existing ecosystems, can have a substantial effect on wildlife. This may be depressive, as for instance is especially true in the case of eucalyptus or gum tree plantations where the gum trees themselves are poorly frequented by animals. Relatively few species of birds breed in them although they are used as roosts by birds and fruit bats.

Introduced animals

Compared to the number of introduced plants that of introduced animals is not particularly large in Africa.

So far none of the introduced mammal species has spread widely and assumed pest proportions on a large scale. The American gray squirrel (*Sciurus carolinensis*) was imported into the Cape Province of South Africa probably early in the present century. The species now occupies an area within forty miles of Capetown (Davis, 1950). This squirrel has become a nuisance to fruit growers and forest plantations because of the damage it causes.

The nutria (*Myocastor coypus*), a native of South America, has established a colony on the Soleo Ranch in Kenya and has also been liberated into the wilds in Zambia. Fallow deer (*Dama dama*) have been introduced on some estates in South Africa, and Ellerman & Morrison-Scott (1951) speak of the presence of this species in North Africa. The wild boar (*Sus scrofa*), introduced into South Africa in the 1890's, now occurs in forest between Port Elisabeth and Cape Town (G. Schürholz, *pers. comm.*).

Although so far only few exotic fish species have established themselves in African freshwater environments, it is clear from the few examples that are available that their introduction can lead to the entire or partial elimination of indigenous species. The introduction of trout in South Africa has apparently resulted in the extinction of an indigenous cyprinid fish. Populations of common carps, largemouth and smallmouth bass no doubt have had a depressive effect on indigenous species. The common carp (*Cyprinus carpio*), several trout species (*Salmo trutta*, *S. gairdneri*, and *Salvelinus fontinalis*) and two species of bass (*Micropterus dolomieni* and *M. salmoides*) have become well established in suitable waters in southern Africa. Trout and bass species are also present in East Africa. Unfortunately some of these species have had a marked and deleterious impact on some river systems and the indigenous fish fauna.

The preservation of endangered species

The problem of the extinction of species was until recently not taken too seriously, even by biologists, because we had been living in a world in which 'new frontiers' and discoveries were always before us in the plant and animal kingdom. This is now no longer the case for most species have been described and indeed unfortunately the total number of described species is now steadily on the decrease as the processes which bring about extinction accelerate. This is particularly true of those forms of greater interest to man: the larger mammals and birds among the vertebrates and the showy orchids and beautiful ferns among the plants.

It is only recently that people have begun to realize the consequences of extinction and in some countries concerted efforts are being made to halt the process and to protect those species already on the 'endangered' list of the International Union for the Conservation of Nature and Natural Resources. But this is easier said than done and requires not only adequate law enforcement but also specific protection of the habitats of the vanishing species. This sometimes proves a virtual impossibility because the last refuges of these species may be on fertile land, much needed by man for his agricultural pursuits.

Species of animals and plants come to the point of extermination in two major ways; either directly through killing or indirectly through the changing of their habitat. The latter approach is of course less obvious than the former, but often contains a more widespread and dangerous threat to survival, which is normally dependent on the habitat remaining intact or perhaps slightly modified. If the ecological conditions change too violently survival is no longer possible.

Human activities of cultivation, flooding and the construction of cities have virtually annihilated entire habitats, resulting in serious or total depletion of the flora and fauna. This is, for example, the case in the Nile delta. The extent and degree of this change are usually in direct proportion to the numbers of humans involved. Consequently, as human populations expand, a steadily increasing number of animals and plant species are being literally squeezed out of existence and will continue to be so.

Less important, but nevertheless effective causes of extinction are introduced animals, particularly imported predators. For example, the Indian mongoose was introduced into the West Indies to exterminate the already introduced Norway rat, but instead of doing so, it managed to eliminate several indigenous mammals and birds and to depress the numbers of other species. In an ever-increasing number of cases, the only lasting way to preserve a threatened species and its habitat is within a national park or equivalent protected area. To accomplish this objective, the protected area must be large enough to accomodate the various habitats of the species, and in the case of larger migratory ungulates, the area involved should provide the necessary food and other habitat requirements throughout the yearly cycle, otherwise the animals are liable to become poached out when they are forced to move outside this protected zone.

The question of survival of threatened species is of international concern rather than a national or local problem. After all, the distribution of species has nothing to do with the establishment of existing political boundaries and when a species does become extinct, the irreparable loss should be everybody's concern. This problem is therefore of urgent international priority within which national parks or equivalent reserves play an important role. This has been well recognized by the World Wildlife Fund in regard to its various activities of funding projects, but it is the countries themselves which have a definite interest in protection of their wildlife and these therefore should give priority consideration to the problem.

Unfortunately, in many developing countries this is not the case. The abundance of wildlife in nature is a source of profound aesthetic satisfaction to many people. To reduce this abundance in man's environment is to invite instability of biotic communities, to restrict man's freedom to choose new species for domestication or other uses and, of tremendous importance, to impoverish the quality of his life.

The problem of endangered species in Africa is less well known than for instance in North America where it has been studied in considerable detail. This is probably due to the fact that Africa has such a wealth of wild animals that people may overlook that the process of extinction is also going on in the Dark Continent.

Wildlife populations have decreased considerably in Africa in recent decades and this process is still going on. Unfortunately it is mainly the large mammals which are most endangered, usually due to destruction of habitats. Compared with other continents, fortunately few forms of African vertebrates have actually become extinct. Among the mammals these include two subspecies of lion and two subspecies of hartebeest, the bluebuck (*Hippotragus leucophaeus*), the rufous gazelle (*Gazella rufina*), the quagga (*Equus quagga quagga*) and the Burchell zebra (*Equus quagga burchelli*). This extinction of certain species is so far most marked in both northern and southern Africa.

The list of endangered species is unfortunately much longer than that of extinct species. Not less than 14 species or subspecies of mammals and four species or subspecies of birds are threatened (see table 2). If effective action is not taken soon to protect these forms, I am afraid that some of them will join the ranks of the extinct in the not too distant future.

It is not generally realized that the total or virtual elimination of certain species can cause imbalances in an ecosystem. An example is the virtual elimination of the leopard from large areas of Africa, resulting from persecution for its much wanted fur. This destruction has caused considerable local proliferation of wild pigs, baboons and other monkeys, which have caused much damage to crops. It should therefore be clear that the protection of endangered species is not only important from an aesthetic and scientific point of view, but also for the preservation of well-balanced ecosystems.

Table 2. Endangered species in Africa

Mammals	causes for disappearance
Tana river mangabey (*Cercocebus galeritus*)	habitat destruction
Tana river red colobus (*Colobus badius rufomitratus*)	habitat destruction

Mountain gorilla (*Gorilla gorilla beringei*)	habitat destruction, hunting
Barbary leopard (*Panthera pardus panthera*)	habitat destruction, hunting
Nubian wild ass (*Equus asinus africanus*)	interbreeding with tame donkeys
Somali wild ass (*Equus asinus somalicus*)	habitat deterioration
Mountain zebra (*Equus zebra*)	erection of game proof fences
Giant sable antelope (*Hippotragus niger variani*)	limited range
Scimilar-horned oryx (*Oryx tao*)	overhunting
Addax (*Addax nasomaculatus*)	overhunting
Swayne's hartebeest (*Alcelaphus buselaphus swaynei*)	overhunting
Slender-horned gazelle (*Gazella leptoceros*)	overhunting
Beira (a gazelle) (*Dorcatragus megalotis*)	overhunting
Walia ibex (*Capra walie*)	overhunting
Birds	
African lammergeyer (*Gypaetus barbatus meridionalis*)	poisons
Prince Ruspoli's touraco (*Tauraco ruspolii*)	restricted range
Dappled bulbul (*Phyllastrephus orostruthus*)	restricted range
Teita olive trush (*Turdus helleri*)	restricted range

Source: The red book. Wildlife in danger. FISHER, J. *et al.* Collins, 1969, 368 pp.

The need for preservation of natural vegetation

One of the most important effects man has had on the African environment through his modification of the vegetation is his eliminating, or virtual elimination, of many species of plants and animals. There is no doubt that this process is continuing at an accelerating pace. Is this something that we can allow to continue without serious detriment to the environment?

Mankind has benefited greatly from species diversity because this has offered him a great variety of wild plants and animals from which to choose those most useful to him, including domesticated varieties. But unless substantial areas of the earth's surface are rapidly preserved from further exploitation and despoliation, we can only foresee the disappearance of increasing numbers of species.

Five important reasons can be given to support the need for conservation of natural vegetation. These are: (1) present economic potential, (2) possible future economic potential, (3) soil and water conservation, (4) scenic and recreational values and (5) habitat for animals which would otherwise disappear.

It would seem rather short-sighted to assert that the majority of species existing, but not efficiently used today, in all probability hold no potential for the future: New products are continually being developed from seaweeds, forest and range plants, and fungi. A continuous search by plant collectors is going on for new drugs or chemicals derived from plant sources. Biological screening of large numbers of plants frequently brings to light new chemical structures that would otherwise not be readily discovered.

It is increasingly recognized that natural plant communities serve as gene pools for the conservation of plant material that may hold future medicinal, nutritional or economic potential.

A substantial population of each plant species is required to preserve not only the species as such, but also a range of variation within the species. The actual number of specimens required varies however greatly between species. It is clear that when thousands of species are involved as in the case in Africa, the idea of keeping mere samples in a botanical garden is quite insufficient as the only way to conserve these species adequately is to maintain a sufficient number of samples of natural forest and other vegetation types. Tracts of natural vegetation are also required for the preservation of animals; in fact, large, wide-ranging ungulates need vast areas in order to survive.

Although there has been an ever increasing awareness among scientists of the danger facing us by depletion of gene pools, this view unfortunately is not generally shared by the land managers, be they farmers or foresters.

Although particularly the latter should know better, they often see little point in protecting an unproductive forest for the sake of preserving a few rare species of plants. More educational efforts along these lines will therefore be required in the immediate future.

The scenic value of natural vegetation is increasingly recognized, and many picturesque places owe their beauty largely to this vegetation. For example, the alpine areas of Africa are extremely beautiful and are therefore an added tourist attraction. As areas covered with natural vegetation become increasingly rarer, more tourists are already visiting interesting botanical reserves in addition to their usual visists to wildlife areas and I predict that this will occur more frequently in the future.

I hope that the foregoing has made it clear that natural plants do have many values to man, some of which are of great importance and may become even greater in the future.

The special need for forest reserves

It seems likely that most of the natural tropical rain forest with the exception of forest reserves, will be destroyed before the end of the century. This is a considerable loss from a scientific point of view, because this forest is a very rich plant and animal community that includes many of the most beautiful and bizarre forms of life.

It is also plausible because of the scarcity of timber that more and more pressure will be exerted on the governments concerned to utilize even the forest reserves themselves for commercial timber production.

If only selective logging took place, the composition and character of these reserves would still be bound to change. It is therefore high time to take stock of the value of the maintenance of natural vegetation in these reserves before further irreversible destruction takes place. Tropical rain forests have much research value because they offer an insight into the complex principles of ecological balance and into the processes of evolution.

The value we put on different trees and flowering plants may in fact change. A weed species, or something we eradicate today, may turn out to be very useful in the future. A good example of this is the timber tree *Afromosia etala* which was virtually unknown before World War II and has now become a very desirable species for the timber trade.

The other great value which I referred to before and which I am sure will become more prominent, is the attraction of natural forests to tourists who will want to experience the 'strangeness' of a tropical forest or the wilderness value of a less exotic one. Seemingly forest reserves will soon be the only such places left, and who can now predict what psychological value these havens of peace away from the modern technologically thinking world will have in the future!

The environmental values of forests

Apart from its role as a source of wood, the forest has also social and protective functions, and affects the climate to some extent.

The influence of forests on the climate is well known, but has unfortunately been inadequately quantified in Africa. The rainforests of West Africa are responsible for replenishing the south-westerly winds with moisture, without which the savannas would have less rain, lower relative humidity and a shorter rainy season (Molski, 1966). Rakhmanov (1966) has documented how the amount of precipitation increases under the influence of forest. As the total duration of the rainy season is the most important factor upon which the natural vegetation depends, the consequences of clearing of tropical rain-forests could be rather catastrophic as there are indications that the micro-climate has changed and the precipitation is reduced in the adjacent Sudanian zone (Molski, *op. cit.*).

The value of the vegetation for the maintenance of water supplies and prevention of soil erosion is generally well recognized and most governments now make efforts to maintain protecting forests. Natural tall montane forest and bamboo thickets serve a most useful purpose of storing infiltrated water and controlling stream-flow or surface run-off.

The removal of forests greatly affects the drainage pattern and sometimes causes the streams to dry up. The process of forest removal may be irreversible, e.g. when soils are shallow and erodable.

On the steeply sloping land of mountain areas and also in the tropics generally a certain plant cover is needed to preserve the fertility of the soil. In some parts of the continent, heedless destruction of the forest in the past has already led to severe erosion of the soil. Also, indiscriminate felling and forest fires are heightening the danger of erosion.

The environmental values of tree plantations include that they protect the soil and serve as shelterbelts.

The need for national parks and equivalent reserves

The national parks movement has grown tremendously since 1872 when some farsighted people suggested and implemented the establishment of the world's first national park in Yellowstone, Wyoming. Each year the number of visitors to national parks in the United States amounts to millions and some of the parks have become so popular that serious consideration is being given to limiting the numbers of visitors in order to avoid excessive damage to the environment.

This interest has reflected very noticeably on national park development in Africa and by far the greater majority of African countries now have one or more national parks. Some, like the Serengeti national park in Tanzania, or the Krueger national park in South Africa have already become world famous and it will only be a matter of time before they will also suffer from visitor congestion, already a problem on certain weekends in some parks, like the Nairobi National Park.

National parks have many values, the main ones being the provision of much needed recreation away from congested areas in unspoiled, natural surroundings; the protection of biological reserves where gene pools of flora and fauna may be maintained which may ultimately be of great value to man; and finally the considerable economic benefits which result from visits and services rendered to visitors.

As our environment is being constantly modified, national parks are now almost the only places where the evolution of ecosystems untampered by man can be studied and they therefore provide significant data on the productivity of natural areas.

It is only during recent decades that African Governments have become aware of the particular value of national parks as a form of land use, and of the need for setting aside those parks or equivalent reserves as representative samples of natural ecosystems within their boundaries. However, the value of national parks as tourist attractions has been well recognized for some time and the national parks of East and South Africa with their magnificent display of wildlife have proved real drawing cards for international tourism.

In East Africa, where until recently the emphasis has been on parks for plains game, several mountain tops have been declared national parks and are visited by many tourists who are interested in seeing the strange flora which has developed on them.

Perhaps in Africa too much emphasis has been placed on the need for establishing national parks rather than nature reserves or wildlife refuges. At any rate, there are many areas which do not meet the internationally established minimum standards for national parks which should be given adequate protection because of rare, endangered or spectacular plants and animals which are to be found there. The establishment of such sanctuaries needs to be given special attention if adequate protection is to be provided to floral and faunal resources of importance to the biology and welfare of mankind. It is particularly essential that at least nature reserves be established on those lands valuable to agricultural production, since they will protect certain species or ecosystems unique for those lands!

I have described (DE VOS, 1969) some of the difficulties of protecting nature sanctuaries. Because they are generally smaller in size than national parks, they need to be more carefully guarded against human destruction. Certain species are particularly vulnerable to changes in their habitat. For example, when natural forests in the Cameroon were opened up for logging operations, the loss in humidity alone was sufficient to cause the death of orchids there (SANFORD, 1970). Many other examples could be cited of the vulnerability of such sanctuaries to the infractions of human beings.

It is not only the national parks themselves that require careful protection and management but the areas surrounding these parks should be limited to use of the land for certain purposes only. This is more particularly the case if the park animals concerned move outside the park during part of their life cycle and for this reason 'buffer zones' are often established around these parks. There is bound to be an increasing problem in these as the human population increases and very careful supervision of the land use in these zones will become a necessity.

Where intensive agriculture close to park boundaries is unavoidable, the construction of fences or moats may be necessary. Moats, if covered with vegetation, are very effective barriers. Even elephants are prevented from crossing them in the Aberdares National Park in Kenya (CLARKE, 1968).

Animal influences on the grassland environment

Grasslands, the natural grazing resource for herbivorous animals, may produce many palatable plants which provide good grazing along with others which do not because they are either coarse, poisonous or both.

Grazing is a most important modifying force in the grassland environment as the degree of grazing pressure affects the subsequent yield and composition of the range. In addition to eating the vegetation, animals trample and cut the soil with their hoofs. Photosynthetic processes are of course reduced if the green leaves are removed, and damage to the growing stems affects the reproduction of the species. Lower growing plants tend to increase as the taller species are usually grazed first. It can be seen from the foregoing that the evolution of grassland communities must have proceeded under the continuous influence of herbivorous ungulates. Grazing succession on African rangeland is usually three tiered: heavy animals like the elephant, hippo and buffalo do the initial trampling of the rank grass; then large herds of medium weight animals such as topi and zebra follow and maintain the short-grass stubble in a favourable condition during the dry season. Finally, light animals such as the kob and reedbuck concentrate on the short grass lawns, and thus a well organized mosaic is developed.

Localised heavy trampling by wild ungulates may cause so much denudation of the vegetation that erosion sets in, creating small hollows. These hollows are often enlarged by the pawing and wallowing activities of many species of wildlife, so that eventually they hold water and are a very useful addition as a source of water for the herbivores.

Some species of herbivores have a particularly dominant influence on the environment.

The elephant is the prime example which by mere bulk is capable of changing entire habitats. He achieves this by digging water holes, pushing over trees or barking them – the former resulting in immediate death, the latter in a slow one – and thus opening up forest stands which subsequently become more habitable due to the increased herbacious vegetation for plains ungulates like Grant's gazelles and zebras. Even elephant defaecations are noteworthy from the point of view of spreading seeds. The buffalo, another dominant animal, in addition to trampling the vegetation aids germination by pushing seeds into the ground with its hooves.

Many small mammals like hares, springhares, ground squirrels and various other species of rodents may under certain conditions build up to such numbers that they drastically modify both the vegetation and carrying capacity of the range. However, grazing mammals are not alone in affecting the vegetation profoundly, for various species of birds, ants, crickets, grasshoppers and termites (see p.145) can also have a significant impact when they exist in sufficiently large numbers.

Frequently, invertebrates become so numerous that serious grazing capacity loss results. For example, when swarms of grasshoppers invade a new area they are capable of removing the remaining herbacous vegetation in a matter of hours. Insects commonly attack range plants in the greatest numbers when drought conditions prevail, and the effects of their activity are therefore very noticeable.

The role of termites and termitaria

Many invertebrates such as locusts, ants and termites exert pronounced effects on the environment, but among these the last group leaves the most obvious evidence of its activities. Termites or 'white ants' are social insects that may either spend all their lives in underground chambers, or they may build mounds on the surface. When they forage, they build tunnels so that they are not exposed to light. Termites are extremely important soil insects as they break down wood and other organic matter. Some species have protozoa in their hind gut to help them digest the wood. Others eat soil containing vegetational debris. A third group forms a working partnership with fungi.

Termite mounds or termitaria are a typical feature in most African landscapes. They may attain considerable height – sometimes up to 10 m – and diameter, and they are made by comparatively few species. The large termite mounds that can be so frequently seen in central Africa are mainly made by non-fossil species. Gourou (1966) found that in the Katanga (Zaïre) three giant termite mounds occur to the acre and occupied 6 per cent of the area. Each species of termite will build its own specific mound.

Termites are a dominant component of the insect fauna of the African tropics, and particularly of the savanna. On certain soils they may well be responsible for the greater part of such aeration and water penetration as does occur. Termite activity has many negative repercussions in savannas, particularly on account of their attacks on young trees which may result in death. However, this activity also can be beneficial by thinning over-stocked stands of trees.

While living termitaria are rarely covered by vegetation, uninhabited termite mounds are often covered by dense clumps of woody vegetation, and on open savannas shrub and tree growth is often almost entirely restricted to these termite mounds.

Since termites carry vast quantities of dead vegetation into the underground galleries of their mounds, these mounds may become islands of rich soils in the otherwise rather barren savanna. In active termite mounds damp soil is brought to the surface from deep in the termitaria where it is displaced in the process of subterranean gallery construction. This soil provides a favourable site for the germination of seeds, many of which have been carried there by the termites themselves or by birds or other animals that like to sit or stand on the top of these mounds.

Because termites continuously bring organic matter to their mounds, these may have a higher percentage of humus. They may also have a higher water retention capacity. With regard to mineral content, it seems that this is richer if the subsoil brought up is richer and poorer if the subsoil is poorer.

A study of the effect of the large mounds of Macrotermes on African agricultural soils has been made by Hesse (1955) in East Africa. He found that plant nutrients or plant toxins were not concentrated and that mounds simply consisted of subsoil. The superior growth of certain plants may be ascribed to better drainage or to accretions of calcium carbonate when drainage is impeded. In some places, the principal arable land is located on flat-topped termite mounds, about one meter high. For instance, in the southern Congo and northern Zambia it is a strange sight to see maize growing on termite mounds, while the level land around is covered with trees.

Termitaria are used by mammals for various purposes, for instance by antelopes as salt licks. Bushmen set traps and snares around termite holes in order to capture mammals such as springbok which are attracted by these

Photo 22. Termitaria are a frequent occurrence in savannas. Termitarium overgrown by shrubs. Luangwa Valley, Zambia. Photo: A. DE VOS.

deposits. Termitaria also are utilised by some antelopes in their territorial activities. Many other animals, such as aardvark, use them for their burrows which in turn are being used by numerous other vertebrates such as warthog, hyaena, lizards and snakes. Foraging by termites is usually selective in nature, radiating in a circular manner from the mounds and resulting in bare patches of ground surrounding the mounds and these serve an excellent purpose of protecting vegetation growing on them from the devastations of fire.

Many termites feed almost exclusively on grasses. Although inadequate information is available about the exact role of termites in the destruction of grasses and of plant residues, enough is known to state with certainty that most termites aggravate the effects of over-grazing around their mounds. Clearly, termites can do a lot of damage to man by eating trees, forage and certain crops.

The role of the tsetse fly: Africa's boon or bane?

Much has been written about the role of the tsetse fly (*Glossina* sp.) as a scourge to man and beast on the one hand and the protection of bush environments against overuse by man and his stock on the other hand. In 'fly belts', or areas infested by tsetse, man has modified the environment far less than in tsetse free areas because the risk was too great that either he himself or his livestock would die from trypanosome infections.

Tsetse are blood-sucking flies restricted to tropical Africa. These flies might well be rated as the largest single insect problem in Africa, since the presence of tsetse has virtually eliminated wide tracts of potential ranching land for use by livestock and has precluded the development of mixed farming in some regions. It is a remarkable example of how a small group of insects have had a major influence on the African environment. Through their being the major vector of human and animal trypanosomiasis the various species of tsetse have profoundly influenced about 10 million km^2 of tropical Africa and have acted for a long time as a buffer against man's encroachment on African biotopes. Nagana is the collective name given to infections by three kinds of trypanosomes passed on by tsetse to cattle. Although under favourable conditions cattle can survive an acute infection of this disease, and pass into a chronic stage, they will not survive under stress conditions. Some breeds of cattle, when infected, are considerably more resistant to the effect of nagana than others: the N'dama and other west African dwarf breeds demonstrate such resistance, but unfortunately, these breeds mature slowly.

It has been estimated that if tsetse could be controlled, the cattle population of Africa could be more than doubled. Were it not for nagana, both the meat and milk yields of African cattle could be enormously improved by crossing with European breeds. The latter are, however, highly susceptible to trypanosomiasis. In addition, the absence or scarcity of cattle in huge areas of Africa results in a lack of manure for the fields, which in turn, adversely affects the maintenance of fertility of agricultural land.

Photo 23. Sleeping sickness is a wasting human disease which also attacks domestic animals and is endemic in certain parts of Africa. Pathfinders picking up tsetse fly specimens in Botswana. WHO photo issued by FAO.

All species of tsetse are closely associated with forest, trees or bush, at least as breeding places. The control of tsetse by altering vegetation represents an indirect attack on the trypanosome by destroying its vectors. As the struggle against tsetse cannot fail to include efforts to replace woody vegetation by grass, there is an essential conflict between the desire to control tsetse and the desire to preserve the forest.

Large areas in Africa have been cleared of bush cover in order to eliminate tsetse fly infestations, but unfortunately, many of these areas subsequently have not been used for either intensive livestock raising or for agriculture in such a way that further bush encroachment could be kept under control. Consequently, after the bush became re-established, the tsetse returned, and considerable funds and efforts were wasted.

A controversy of long standing has raged over the need to eliminate wildlife as part of effective tsetse control. The rationale behind eliminating game animals is that the flies are denied access to a source of blood: if wildlife can be eliminated from a fly-infested area and if that area is subsequently maintained free of game, the fly should be unable to live there. The practice and policy of eliminating wildlife by hunting has been reviewed by GLOVER

Photo 24. Sleeping sickness is carried by a species of bloodsucking fly called 'tsetse'. Close up of a tsetse fly. They range in length from 6 to 16 mm. WHO photo issued by FAO.

(1965), who concluded that there is no valid justification for game destruction as a practical or lasting means of tsetse control, because (a) it is probably impossible to exterminate smaller ungulates and (b) larger animals continue to migrate into the shooting areas. Although it is now more generally recognized that complete elimination of wildlife is not necessary for effective control, hundreds of thousands of animals have been unnecessarily slaughtered in several countries. in attempts to control tsetse. As a result, many species of game have, for example, been completely eradicated from large areas in northern Uganda. This is one of the most tragic examples of how agencies charged with tsetse control never had the resources or took time to investigate the wider ecological implications of their activities.

The present approach in reducing the numbers of game in order to eliminate tsetse is to try to reduce as much as possible such preferred hosts as warthog, kudu, bushpig and bushbuck. Other species, such as waterbuck, hartebeest and zebra, which are rarely used by tsetse flies for blood meals, can be left unharmed. Clearly many mistakes have been made in tsetse control. When further plans are contemplated for the opening of fly belts, more attention should be given to the need for evaluating the use of wildlife and forest resources for their economic and other assets. Before management plans for fly belt areas be implemented and these belts be cleared of flies, much more information should be collected on the value of these existing resources.

Today we witness considerable efforts being made and vast amounts of money being spent to keep settlements and grazing lands free of tsetse. These efforts will no doubt be accelerated, because with the increase of the indigenous population and its livestock, the role of the tsetse fly as the guardian of environments more or less untouched by man is coming to its end.

Better methods of tsetse control will eliminate these insects from areas where until now this has been impossible. These methods include ground spraying with insecticides, which has been used with considerable success (Lambrecht, 1972) and the sterile male technique, which is still in an experimental stage.

It is to be hoped that future tsetse eradication schemes will be based on well planned land use programmes and that many of the mistakes that have been made in the past will be avoided.

The ecology and control of the desert locust

Probably the most damaging invertebrates to range and croplands are grasshoppers, locusts and crickets. The devastations of locusts in Africa are often considerable.

One of the best-known, and potentially the most dangerous genus of international pest is the swarming locust. The term 'locust' covers a number of species among which the desert locust (*Schistocerca gregaria*) and the red locust (*Nomadacris septemfasciata*) are the two most damaging to man's agricultural economy.

The reciprocal interactions between man's modifications of his environment and the population dynamics of the desert locust provide an interesting insight into the ways by which man can affect the ecology of a species. Desert locust swarms invade extensive areas in north, west and east Africa, while red locust swarms are mainly restricted to south and central Africa. Invasion areas of the two species overlap in East Africa.

The desert locust is a nomadic species in Africa and western Asia. Under certain conditions enormous swarms are produced and carried by prevailing winds to areas where they can do substantial damage to crops. The area which is usually invaded by these swarms is the largest known for all species of locusts (LOWE, 1970). Before control practices were initiated this locust was one of the most serious plant pests in Africa.

The successful control that has been obtained over the desert locust over several decades is largely the result of extensive ecological research. The species has been found to have extreme ecological adaptability. For instance, no permanent breeding areas exist for this species. Considerable problems remain to be solved however, and these may well be intensified in the future because of the need for continued rapid development of agriculture. Ecological research has produced reasonably efficient and economical chemical control measures, without untoward biological side effects, and may in time produce an acceptable permanent solution, although this is not yet foreseeable.

The most serious problem connected with locust control is the effect of development projects, particularly agricultural schemes. The incidence of locusts and similar pests is often directly related to particular ecological disruptions made by such projects. Indeed, locust swarms are sometimes a man-induced scourge (HASKELL, 1972), because man creates more favourable conditions which enable this insect to complete its life cycle. Increases of local water supplies in the desert favour the locust because fresh vegetation is required for development from the egg stage to maturity. Moreover, locusts quickly become pests of grain crops when these become available and also increase in abundance on overgrazed pastures (LOWE, 1970).

The development of irrigation projects has resulted in sudden increases and has involved seasonally massive build-ups of locusts. Non-desert species of locusts are aided in crossing deserts by man-made irrigation canals as well as by natural rivers, and by 'oasis-hopping'. A hazard of oasis cultivation is that of presenting targets for nomadic and migrant insects such as desert locust swarms which frequently descend upon the oases (UVAROV, 1962).

On the subtropical edge of the Sahara disturbance by cultivation has recently caused swarm-breeding. Desert locusts, as well as many other species of locusts and grasshoppers, quickly become threats to grain crops produced on newly established farms. They also become pests under conditions of overgrazing and other land malpractices (UVAROV, 1962).

Photo 26. Newly-fledged adult desert locusts and a fifth instar locust (near bottom) on a palm leaf. This species has population outbreaks during certain years which can be very devastating to crops. Photo: FAO.

Photo 25. More than 40 countries, from West Africa to North India, are threatened with desert ◁ locust outbreaks. International efforts to control this locust are being co-ordinated by FAO. Locust swarm in Ethiopia. Photo: FAO.

Much has already been written about the maligned goat. One of the more interesting papers about it is written by FURON (1958), entitled: 'The gentle little goat, archdespoiler of the earth', in which he points out that the goat is one of the worst enemies of mankind, because it does not only browse and graze almost anything edible within reach, but even pulls out plants and climbs into the overhanging branches of certain trees. He draws the conclusion that the goat represents a grave danger and that it is the cause of the disappearance of vegetation and serious erosion. No doubt most of those who have seen goat herds in action in various parts of Africa will very much agree with this statement: there are a great many overbrowsed and overgrazed areas where the forage is too poor for cattle and sheep to continue to survive, and where goats are engaged in polishing off the remainder of the vegetation. There are also examples, for instance from St. Helena, where goats have destroyed forested habitats after their introduction in a matter of a few decades.

The question still remains whether or not the goat is detrimental to the range under all circumstances and whether in fact the goat has normally initiated the serious range conditions that are so prevalent wherever it is kept. Some range scientists are now of the opinion that the goat normally only commits its ravages after man has already started damaging the range through fire or overgrazing by cattle and that the goat is only doing its damage at a secondary level and is not the 'causative agent' for the problems observed. Perhaps this is somewhat of an academic argument, but it is certainly worth considering from an ecological point of view.

Finally, it must be admitted that under good management the goat can be, and is, valuable animal to the Africans: it is extremely adaptable, disease resistant and produces good meat, milk and cheese. Animal scientists therefore in general advocate the introduction of improved breeds of goats, provided that their herds are properly managed.

Wetlands, estuaries and mangrove swamps

African wetlands are being drained, burned, grazed or otherwise modified at an accelerating pace. In addition, many wetlands probably have disappeared as a result of a lowering water table. Several place names in the Republic of South Africa ending with 'fontein', the Dutch word for spring, were given this name due to their adjacent wetlands or springs which have long since disappeared, due to the lowering of the water table.

Wetlands are important in Africa for a variety of reasons. They offer grazing to livestock and wildlife during periods of drought; they provide habitat for a variety of fish and invertebrates; they provide resting and nesting places for a great variety of wildfowl and other aquatic birds, and mammals, such as the sitatunga, which prefer this type of habitat. Wetlands may also,

Photo 27. Goats can be very destructive to already overused ranges. A hungry goat stripping what little green is left on thorn trees in northern Upper Volta. Photo: FAO.

when drained, be among the most productive agricultural land available. It is not surprising therefore, that with a steadily increasing human population, more and more wetlands are being drained and converted into agricultural land. Some of the most productive wetlands from a floristic and faunistic point of view already have been drained and now produce substantial quantities of rice and other important crops.

Many tidal and estuarine swamps are being converted into rice growing areas with multilateral and bilateral assistance because rice nutritionally is much superior to yams and cassava. But these swamps also are, or were, among the most productive natural biotic communities to be found in Africa. These are areas where large numbers of wildlife rest or nest, and where the dugong, or seacow, finds a last refuge. This species has become rare along both the west and east coast of the continent.

The drainage of wetlands is not always as successful as originally expected.

Often it is problematical whether the considerable capital investment necessary can be maintained and whether the potential fertility of the soils can make such investments profitable. Often capital outlays are higher than projected and fertility decreases after some years of exploitation. Generally, however, the carbohydrate grain production of a drained wetland exceeds the protein production that could be obtained under natural conditions; certainly more mouths can be fed. In addition to being drained, African wetlands also are being flooded to aid in the production of hydro-electric power.

The Kafue Flats in Zambia, once one of the richest wildlife areas in southern Africa, have recently been flooded, threatening the survival of the red lechwe (see p. 83) and also one of the most spectacular displays of aquatic bird life to be seen anywhere. Another potential threat to wetlands is pollution. Although pollution via pesticides and fertilizers is still a relatively rare phenomenon in Africa, there is no question that wetlands pollution is becoming more widespread as the need for greater agricultural production becomes more urgent. Pollution by artificial fertilizers already presents local problems and pollution of estuaries by raw sewage from cities further upstream is becoming an increasing problem in the Ivory Coast, Ghana and Nigeria.

To date, the mangrove forest areas along the west and east coasts of Africa have not been exploited significantly. They are used to some extent for fuel and small construction timber. The destruction of these forests would remove a major source of food for many wild animals and therefore reduce the productivity of the shallow waters. Unquestionably the need will arise for a more intensive exploitation of these forests, as already has been the case in southern Somalie, where mangrove swamps have virtually disappeared.

Many estuaries and wetlands are highly productive of protein in the shape of animal flesh – mammals, birds, fish and invertebrates. The aesthetic loss of many beautiful species of birds, due to the drainage of wetlands, is considerable.

As only incomplete surveys have been made of the remaining unmodified wetlands in Africa, it is difficult to say how many more can be drained or otherwise affected by man without causing irreparable loss to these valuable biotic communities, but indications are that not many more should be drained and that in fact remaining wetland and mangrove areas should be carefully protected and managed. Some wetland areas are now protected as part of a national park or reserve system, but the proportion is insufficient and many more should be added. Some wetlands and estuaries should be set aside for the special protection of endangered species, such as red lechwe and the sea-cow, or for rare or disappearing species such as the shoe-billed stork and the wattled crane.

Finally, it should be realized that an integral part of any policy of wetlands protection is an effective system of enforcing legal protective measures. Fire in particular is a constant threat to wetlands because local people increasingly are inclined to burn the margins of wetlands to obtain new vegetative growth for their livestock during periods of drought. It will be most difficult to prevent these and similar activities destructive to wetlands without the continuous presence of adequately trained protection staff.

The pesticide problem

Environmental problems caused by the application of pesticides – insecticides, herbicides, rodenticides and fungicides – can become serious in Africa, as increasing amounts are being applied. Fortunately so far in most parts of Africa the use of pesticides has been largely confined to cash crops for export, and to disease control.

Two main groups of synthetic insecticides can be recognized, the chlorinated hydrocarbons and the organo-phosphates. The first group includes DDT, dieldrin, endrin, chlordane and toxaphene among others. The serious problem posed by the application of insecticides belonging to the first group is that they are not sufficiently specific in their effects: in addition to the intended control of pest species, their predators and/or parasites are often also killed in large numbers, thus inducing increasingly unbalanced conditions in the environment. This is particularly true on extensive plantations, such as cacao, where pesticides have been applied regularly. Many are the instances where control of one pest engenders outbreaks of other pests, and necessitates gradually heavier and more frequent applications of the same or some other insecticides.

This creates economic as well as environmental problems. The chlorinated hydrocarbons are also persistent and therefore accumulate in the bodies of animals, particularly vertebrates who eat poisoned insects and fishes. For example, a concentration of DDT in fish-eating birds endangers their lives and reproductive capacity. Unlike chlorinated hydrocarbons, organophosphates, including malathion, parathion and TEPP, are unstable and non-persistent; thus, they tend not to produce chronic effects in ecosystems or to accumulate in food chains.

Agricultural pests that cause extensive damage are not accidents of nature, but rather fundamental consequences of present agricultural pratices.

A system of monocultural agriculture – extensive areas of one plant crop – provides exceptional opportunities for those insect species whose habitat requirements are provided for by such conditions. Given such large areas of favourable conditions, insect species which under natural conditions would maintain relatively low population densities are able to increase in numbers to pest proportions. Pesticides are applied to reduce these outbreaks, so that crops are not destroyed.

Unfortunately many people in the developing countries of Africa do not realize the hazards associated with pesticides. Applications may therefore result in health problems to local inhabitants or excessive damage to the environment, or both. It is therefore necessary that the application of pesticides be carefully controlled by government legislation, and also that in the future biological control methods be applied together with pesticide applications. The general aim of this 'integrated' approach is to use a mixture of biological and chemical means to keep pest populations down at levels that are not economically significant. Unfortunately, this type of control is relatively expensive and requires careful supervision by well-trained technicians. It can therefore not normally be applied under African conditions.

There are, of course, both short and long term effects of pesticides upon ecosystems. The short term aspect is immediate death of animals of aesthetic or economic value; the long term is the accumulation in soil and water of the residues of insecticides. The latter is particularly dangerous as the effects often become apparent only long after the damage has been done. A serious environmental issue in Africa is the application of insecticides to streams in order to control the black fly (*Simulium*) vectors of onchocerciasis, or river blindness. DDT and toxaphene have been poured into streams to kill the black fly larvae. This control of black flies no doubt has changed ecological conditions in many streams and rivers. The irony of all this is that, aside from the damage to fish, fish food and aquatic organisms in general, considerable damage is done to the most efficient predators of the black flies – stone flies and caddis flies!

Although relatively little experimental work has been done on the effects of insecticides on wildlife in Africa, there are indications that these effects are comparable to those in other parts of the world where they have been studied in more detail. The intensification of agriculture and also increased efforts to control insect vectors of diseases have resulted in the application of thousands of kilograms of thiodan, endrin, dieldrin and DDT in various African countries.

Everaarts *et al.* (1971) state that recent studies in Africa have shown that the use of insecticides can have very important secondary effects among the fauna and that acculumation of these materials has been found in food products, as for example fishes. In their study in cotton growing areas in the Republic of Chad concerning the side-effects of chlorinated hydrocarbon been increasingly fragmented – particularly in densely populated countries insectivorous birds after spraying with endrin/DDT mixtures. The death of birds was mainly due to endrin. Their finding that a few birds were killed by dieldrin showed that the treatment of cotton seed with this pesticide may also give rise to mortality in seed-eating birds. They obtained indications that insectivorous birds already were becoming scarce in the cotton growing areas.

The application of chlorinated hydrocarbon insecticides in tsetse control may give rise to considerable mortality among various species of birds and other vertebrates. Particularly in the drier savannas many forest and game reserves sooner or later will have to face the threat of tsetse control measures.

To conclude, let me state that in my opinion the gravity of the pesticide problem is not yet fully recognized in Africa. Occasionally a report is published about bird or fish kills, but in general the people do not concern themselves much with this problem. I am afraid that some day, because of neglicence, there will be a fairly extensive 'people kill' that will shock everyone. Until then a general attitude of unconcern is likely to prevail.

Land tenure problems

The way the land is held in tenure has a definite effect on the environment. This is certainly true for Africa which is noted for its low level yields and the correspondingly unrewarding small and fragmented nature of the peasant holdings. If this is to change, the average African farmer must of necessity sake modify his tenure of the land. Changes in ownership, land reforms, and changes in the size of land holdings all may create environmental problems.

Most rural people in Africa are influenced by traditional land tenure systems and there are many instances when customary land laws restrict the use of certain areas to conservation purposes only. Traditional systems of cultivation attempted to protect the soil by providing cover crops or fallow periods. On the other hand, most land was held in tribal tenure, which is hardly conducive to conservation practices by individuals. These systems evolved under conditions which are increasingly at variance with present-day requirements and, therefore, revisions are urgently required.

Increasing pressure on the land, the introduction of modern technology together with the development of a market economy are having an increasingly disintegrating effect on the tribal order. The change from abundance to a growing scarity of land has created pressures for alternative land tenure systems. Some of the traditional systems were eliminated by colonial administrators; in many instances attempts were made to introduce the concept of individual ownership. Generally, although family and group ownership of land remains prevalent, changes toward individualization of land tenure are apparent in many areas. The need for increased output while at the same time maintaining soil integrity requires individual tenure. As a result of many centuries of application of inheritance laws to land tenure, land holdings have been increasingly fragmented – particularly in densely populated countries like Uganda, Ruanda and Burundi – to the extent that it has become uneconomical and inefficient to farm individually owned properties. For this reason, land consolidation has become a very important management tool. Although progress towards consolidation has been made, it has been slow and the necessary improvements in the existing land tenure systems are generally overdue. Changes will have to keep in mind existing tribal customs and tabus, and they will also have to respect that security of use of the land is essential, if the individual is to improve current and future production. Preferably any new system should be tried on an experimental basis to assess the reactions of the rural population before it is generally applied. An example of success for land consolidation is given.

Land tenure problems vary greatly between different parts of Africa because of different population densities, crops, soil fertility and tenure systems.

Legislation and adherence to customary law also vary greatly. Land tenure problems are particularly severe in the Sahel and Sudanian zones because of the traditional movements of the nomads. These problems recently have become more severe because these people have built up their herds to larger numbers. Although the countries concerned could potentially control the movements of their nomads, this has not been enforced too much because of the risk of troubles with these people.

The few efforts made so far in improving land tenure so that the risk of land degradation can be reduced have not been very successful. Other efforts made in the humid tropics to regulate the clearing and settlement of land by peasants, whether legal or not, have not been very successful either.

Generally speaking, land consolidation has not been particularly successful in Africa, because either the rural population is opposed to it, or population pressures rather quickly modify gains made. This is no reason for giving up hope.

On the contrary, the speeding up of land consolidation and the improvement of land tenure systems is an absolute must and should be speeded up wherever possible. The Government of Kenya is now making a considerable effort to improve the tenure system on range lands. It still remains to be seen whether the system of land adjudication which they are trying to establish will be successful. Under this scheme communal land is deeded to groups or individuals in the hope that these will improve range and livestock management on their lands. It is unlikely that new legislation can destroy the kinship system and remove the burden it places on successful members of the group. If the practice of landholders to allow landless kin to share their land or till part of it continues, this can only lead to the development of hidden subdivision and further fractioning of the land, while the sale and purchase of arable rights may give rise to increasing fragmentation despite regulations for its prevention (ALLAN, 1965). If land tenure could be made secure, to serve the increase in efficiency, and if the African farmer could be helped toward using more modern methods, a great stride could be made in raising productivity toward higher levels. Therefore, efforts should be continued to change traditional tribal tenure into a system of individual tenure.

Improvements in land tenure systems will necessarily continue to be slow because rapid increases in population pressure are a force favouring further fragmentation of land holdings and politicians are often loath to push for land reform measures against the wishes of the majority of the electorate. What is needed in the first instance is more emphasis on the public relations and education aspect of the problem. The rural population will have to be shown, with examples in their own communities, that land consolidation is a must and to their own economic advantage and that they should accept initiatives by the government in that regard.

V. PROBLEMS, NEEDS AND POTENTIALS IN LAND USE

While in chapter III the main objective was to describe the various effects that man has had on the African environment and to point out how devastating these have been in many ways, in chapter IV an effort was made to inform the reader about some more specific land use and ecological problems to broaden his understanding of the special African conditions.

In chapter V the problems, needs and potentials in the use of the various renewable natural resources will be discussed with the objective of demonstrating that if better use is made of modern technology, if rural people are better trained in land use practices and management procedures, and if more capital can be generated in support of meaningful improvements, assuming that the population explosion can be curtailed, the situation in Africa may not turn out to be hopeless, and that certainly all the encouragement possible should be given, particularly through multilateral and bilateral aid, to move along the various lines suggested. That action is of the utmost urgency needs no further elaboration.

Agriculture

Problems

Rapidly increasing populations and higher expectations of the people lead to an increasing demand for food and this forces farmers to a greater use of natural resources and modern technology. This situation tends to add to the difficulty of ensuring the maintenance of natural resources because new lands are brought into crop production from pasture and range and new areas for livestock production are taken from forest lands. Intensification of crop and livestock production per unit area by the introduction of high-yielding varieties and breeds, and additional inputs such as agro-chemicals also create increasing problems of maintenance of both land and the genetic resources of plants and animals.

Arduous efforts to expand cultivation in the humid tropics have persistently resulted in failure and loss of topsoil. In the Congo basin the Belgians have studied the possibilities of replacing shifting cultivation with intensive cultivation. The results of these studies have been rather discouraging as productivity could only be raised nominally. One reason for this is that when the forest cover was removed, soil temperatures rose dramatically, the humus was degraded, and the plants suffered. The conclusion was reached that only when long, narrow, tilled strips were created in the forests, agriculture could be undertaken for a strictly limited time period.

Progress in agriculture has so far been disappointing. Yields are generally low and much land remains inadequately used. Based on what is now known, the problems of agricultural production will continue to be a severe impediment to progress.

In many countries no satisfactory replacement has so far been found for the bush-fallow system and few cropping systems have been developed as yet which are capable of maintaining high levels of sustained productivity of animal food crops.

In many countries the growth of agricultural production has not kept pace with population increase. In 16 of the 39 countries of developing Africa per caput agricultural production in 1971 was below the 1961–65 average and in several countries, notably Gambia, Guinea, Liberia, Nigeria, Burundi, Mozambique, Rwanda, Somalia, Tanzania and Uganda per caput levels have even shown a clear downward trend (FAO, 1972).

In the least developed African countries, agriculture is a major contributor to GDP, employs a high proportion of the total population and is often important for the trade balance. In Africa as a whole, nearly 75 per cent of the people are still employed in the rural sectors of the economy, mostly at near-subsistence levels of agricultural production (FAO, 1972). This is not surprising because of the lag in development in this continent as compared with the more developed parts of the world and also because of the large proportion of land that is used for agriculture.

Growing pressure on the land has led to the extension of crop farming into areas which, for reasons of limited rainfall, climate, topography or soil quality, are not suitable for such production. Wherever human populations have increased to the point where farmers cannot find forest patches that have not been recently cultivated, they are forced to use recently cropped land, resulting in the exposure of more land for a longer time to torrential rains and finally in much erosion. This situation is all too prevalent in Africa today.

Apart from some food and several commodity crops, Africa has relatively few indigenous plants of economic value as compared to the other continents (see also p. 119). Therefore, progressive food and commodity production depends for the greater part upon exotics which have been imported from other continents. As a result, production has become increasingly diversified.

The increasing commercialization of agriculture which has taken place throughout this century, still continues. Probably the majority of farmers will spend the greater part of their time growing crops for their own use, but the time is now rapidly approaching when efforts will be roughly equally divided between production for subsistence and for sale. In those countries where there are food shortages, such cash crops as groundnuts and cocoa are now increasingly produced for local consumption. Soya is one of the most promising food crops because of its high protein content, the protein of soya beans has the highest nutritive value of any plant protein source.

There are about 180 varieties to pick and choose from in Africa and soya beans can be grown almost anywhere on the continent. However, there are some limitations to production. For instance, in East Africa it doesn't seem to fit in too well in a mixed cropping system, because it takes 120–130 days to mature. This does not leave enough time for a second crop, like sorghum, to grow during the dry season. In addition, soya bean fields require innoculation with nitrogen fixing bacteria whilst ordinary beans mature in 90 days and do not need to be innoculated.

There are various serious impediments to agricultural production which can be alleviated by the use of modern technology. One of these is insect pests, nematodes and vertebrate pests which take a heavy toll of crop and animal production, and of stored crops. Unfortunately, satisfactory control measures are generally not available or too expensive.

As long as heavy losses from plant pests continue to occur, and unimproved varieties of crops are used, the use of commercial fertilizers remains uneconomical.

Losses of agricultural products during harvesting, storage and transport are still very high in Africa and it is vitally necessary to prevent such waste.

One of the first improvement measures for increased agricultural production is the use of better hand implements, preferably manufactured locally where this is possible and justified, while draught animals and farm machinery must also be employed.

The 'Green Revolution'

Until a few years ago, the 'Green Revolution' was considered the most important break-through in agricultural development in this century, so important in fact that it was thought that it would solve the food scarcity crises with which several developing parts of the world would have to struggle for quite some time to come.

The Green Revolution, simple in concept, means modernizing agriculture by getting more yield out of land from high-yielding varieties of rice, wheat and maize as the more important grain-producing crops.

The joy that accompanied the development of these miracle varieties in the 1960's has been tempered by skepticism and disillusionment in the 1970's as famine continues to stalk some poorly developed regions. Although these revolutionary seeds, and the technology that goes with them, still hold the potential for greatly improving the food intake of millions of impoverished people, this potential remains substantially unrealized. The reasons for this are manifold, the most important being that many farmers who switched to the new varieties have not been able to increase their output much, if at all, because they were not able to take full advantage of, or did not have access to, the supporting technology, namely the required fertilizer, insecticides and water.

The new fertilizer – sensitive varieties will require millions of tons of fertilizer, which are not now available. There will also be an increasing dependence on pesticides to control pests among the genetically uniform varieties of grain crops. The poor farmer, living from crop to crop, simply cannot afford the added costs without credit, which is expensive, if available at all.

High-yielding varieties are not yet in common use in tropical Africa, with the exception of maize in East Africa and rice in West Africa. It is not generally realized that these varieties demand much more care in cultivation and up to 50 per cent more in labour requirements than ordinary crops. Such considerations as marketing, storage and transport of the surplus produced are all new problems not faced in the growing of traditional crops. Moreover, some of the high-yielding cereal varieties are already proving susceptible to disease and it is likely that any resistance they presently enjoy will lessen with time.

One by-product of the Green Revolution that its designers perhaps did not realize is that it will increase water pollution problems because these crops require heavy use of fertilizers. Careful attention should be given to this type of environmental damage.

Another unforeseen problem created by the Green Revolution is that it has helped to widen the gap between rich and poor farmers, because those farmers who can afford the requirements necessary to produce bumper crops, become rapidly richer. If only a small fraction of the rural population moves into the modern century while the bulk remains behind, a highly explosive situation is foreseeable.

NEEDS AND POTENTIALS

From what has so far been stated, it seems clear that most of African agricultural and related development has not really been adequately planned, but instead it has grown spasmodically and casually, except for certain plantation crops. In recent decades the more progressive governments have realized that this process can no longer continue in view of the rapidly increasing demands for agricultural products and African countries are increasingly making use of development plans. In these plans stress is laid on rural development and employment, self-sufficiency of food, agricultural diversification and the earning of foreign exchange (FAO, 1972).

Present programmes of agricultural development are rarely preceded in developing countries by an appropriate evaluation of resources and their use. Comprehensive land use planning is often lacking and where it does exist, does not generally involve the local communities.

A developing agriculture must have its own 'infrastructure'. This involves institutions and services which are essential to agricultural development, but which farmers cannot provide for themselves.

The main material input requirements of a developing agriculture fall into two categories. The first comprises machinery, fertilizers, pesticides, herbicides, fencing materials and other products which it is the function of industry to produce. The second is the biological input of new and improved crops and livestock. The rate of progress will be determined by the rate at which trained personnel can be provided for all of the institutions and services required. To step up agricultural production for sale, a certain amount of mechanization will be essential. As PHILLIPS (1959) has pointed out, however, ill-conceived, hasty and inefficient mechanization can do more harm than good. Where mechanization is unlikely to aid to the local practices of production, it should not be encouraged beyond a desirable minimum.

Speed of operation becomes particularly important where planting seasons are short, and new techniques such as weed control and the incorporation of organic matter in the soil will frequently demand mechanization.

Reduced exports of food products are in some countries the result of deliberate policies to divert the benefits of increased output to domestic consumption. Self-sufficiency and import substitution are of particular importance for grains – wheat, maize, rice and sorghum – in certain areas (FAO, 1972). Some countries show a concern for the improvement of nutrition by such means as the partial replacement of cassava by sorghum and millets, extended cultivation of pulses, and increased supplies of animal protein. In an area where meat production per head of population is tending to shrink, pulses may be expected to play an ever-increasing role in human nutrition.

As the fertility of the land declines, there is an increasing tendency among the people to rely more on cassava, a crop which gives high yields even in areas of low rainfall and poor soils and supplies a reasonable amount of carbohydrates and, though poor in proteins and vitamins, is a useful reserve against famine. Only infusions of new technologies of production, which must come from research, will be able to transform traditional to modern agriculture.

It is most difficult to improve or to replace shifting cultivation by another land use system which is less damaging to the land. This can only be done under certain circumstances and gradually because of the low educational standards of the Africans concerned and their lack of access to modern equipment or technology. Shifting cultivation could be improved: the time spent in clearing the forest can be greatly shortened by the use of chain saws and total clearance can be prevented by interplanting forest trees with trees or shrubs of economic value such as oil palms or cocoa. The trend toward the extension of such semipermanent tree crops as bananas and plantains is highly successful.

A relatively obvious method of improving the practice of shifting cultivation consists of substituting the natural fallow by specially planted fallows, perhaps of quick-growing leguminous shrubs, which would have a quicker and better effect on soil improvement than the natural fallow.

Although the use of fertilizers is gradually increasing, inadequate amounts are still used even in the more developed African countries. But fertilizers alone cannot maintain soil productivity and new ways of maintaining soil organic matter must be found if the fallow is to be reduced or abandoned and if accelerated erosion is to be checked. If continuous cultivation is to be adopted, one method or another for maintaining soil organic matter at an optimum level will be necessary.

Future trends

In an increasing number of countries there will be a growing disproportion between population numbers and their food supply, leading to a further deterioration in the levels of living. Farmers will be forced to make greater use of the limited natural resources that are available to them. This situation adds to the many difficulties that already exist in ensuring the maintenance of natural resources. Growing pressure on the land will lead to an extension of crop growing areas: more pasture lands will be taken from livestock production and more forest lands will be converted into crop lands. As this is usually land of marginal value, it is consequently subject to multiple forms of deterioration. More and more agricultural land will also be wasted by non-agricultural activities including the discharge of wastes from urban areas and industries and also the encroachment of urban and industrial areas and transport infrastructures on agricultural lands.

While the most striking alterations in crop patterns in tropical Africa over the past decades concern the spread of the various export crops, to obtain much needed foreign exchange, there have also been some significant changes in the distribution of the crops grown primarly to provide local food supplies. The most widespread change has probably been the expansion of maize growing at the expense of the traditional African grains.

In many west African countries improved cultivation methods and high-yielding varieties are used for establishing and extending rice production. Rice is the main staple in some West African countries and is growing in popularity elsewhere also.

Research priorities include the selection and breeding of crops which are ecologically better adapted to the different zones. Search should continue for improved crops, including bird resistant sorghums and hybrid pearl millet. Intensive studies should be undertaken of the chemical, biological and physical properties of the soils in relation to their utilisation for plant production.

With new advancements in agricultural technology it is likely that new sources of food will be developed and used in Africa. To what extent these will alleviate impending food shortage remains a debatable issue. Some species of yeast, bacteria and algae may become important sources of food. In fact, a microscopic African alga, the *platensis* spirulina, is already gathered in Chad by local inhabitants from alkaline lakes. They dry these algae on sand in the form of cakes which have a high concentration of nontoxic proteins. It seems likely that the food industry will process various species of algae in the future. Various bacteria and yeasts can be selected and cultured to grow on organic wastes. This recycling process could produce much protein.

Much can be done to increase the productive capacity of food crops, but this will require considerable investments in research, an acceleration in the application of technological developments and above all improved understanding of the African farmer of modern land use practices.

Range and pasture management

RANGE MANAGEMENT ON ARID LANDS

Grazing of rangelands is often severly handicapped by climatic difficulties particularly in the more arid lands where water supply is highly irregular due to erratic rainfall. Stock water is consequently often in short supply in the dry periods and the vegetation which dries up often results in wholesale loss of stock. Rangelands tend to deteriorate rapidly under these conditions and in fact deterioration of African rangelands is continuing at an alarming rate, especially in those more arid regions, where a lowering of livestock production has resulted.

Water is obviously an essential need and good range management regarding watering points should be governed by two primary principles: firstly, it is uneconomic for cattle to walk more than 15 km to and from water, and for maximum production this distance should be much less, around five km. Secondly, the number of stock watering at any one point for long periods should be carefully controlled in order to avoid serious overgrazing of the immediate surroundings. So far, little effort has been made by African pastoralists in either direction.

The importance of grazing control cannot be emphasized enough. Even when controlled grazing practices do exist, difficulties often arise with the concentration of animal traffic around the watering points and shade areas. The vegetation is frequently destroyed in these areas and this can and does initiate erosion. As this progresses, the surrounding vegetation is destroyed and the affected part is enlarged. Control and re-stabilization of such areas must be undertaken rapidly before more extensive damage is done. Supplying adequate watering sites and fencing, or otherwise excluding animals from the eroding portions of the pasture can help in overcoming the problem. Unfortunately, many mistakes are made in providing additional water. For example, DUMONT, 1968, p. 151, states that between 1947 and 1961 France spent about 50,000 million old francs on the Sahelian area, in Mauretania, Senegal, Mali, Niger and Chad, to provide water wells and artificial or controlled natural ponds for cattle herds. The areas near these wells became overgrazed and the vegetation suffered accordingly. The points of water supply had been fixed before a map of the pasture lands, which could have established them on a more satisfactory basis, was drawn up.

Certainly the mortality rate of the animals diminished as a result, but there is little point in keeping old cattle, valueless from a protein production point of view, alive within the traditional 'sentimental' framework.

Pastoral people, although not very numerous, own and utilize for their own subsistence the major part of Africa's livestock resources. Although precise data are not available, it seems that there are 60 million pastoralists in Africa, most of whom live in 17 mainly pastoral countries south of the Sahara. These people possess and use for subsistence a livestock resource of perhaps 80 million Livestock Units (1 L.U. = 1 adult bovine or camel, or 10 sheep or goats = about 250 kg liveweight). This source of meat and other animal products is only a fraction of what it could be under proper management and unfortunately contributes too little to alleviate the chronic animal protein shortages.

Improvement of nomadic and transhumant pastoralism in arid and semi-arid areas will prove difficult because most pastoralists refuse to limit their herds to the carrying capacity of the environment.

The digging of new wells in areas where the pasture is inadequately exploited will, in the absence of associated economic and social measures, eventually result in larger herds living under conditions of near starvation and not in an improvement of conditions for the existing herds. In order to prevent this from happening, the appropriate authorities could adopt a system whereby alternate wells are closed, thereby spreading out the pressure on the range.

BROWN (1971) has pointed out that failure to deal with the overgrazing and denudation problems in semi-arid pastoral areas is at least in part due to failure to understand the human population situation and apply remedial measures. Such measures can only be effective if they strike at the root cause of the matter, which is the ecological undesirability of subsisting on milk in a habitat which is ecologically unsuited to milk production. It is necessary, therefore, to substitute something else for milk. This can either be direct dependence on grain itself by cultivation, or it can be through greater dependence on the sale of meat for cash with which to purchase grain. Either involves basic changes in the diet, and indeed in the whole biology and sociology of the pastoral peoples and that is a mighty change indeed!

BROWN (*op. cit.*) has suggested that another method of dealing with the situation could be by the sale of weaned immatures or yearlings to fattening areas. The immatures could be sold off at the beginning of the dry season and thus relieve the pressure on the environment to some extent. However, such an arrangement presupposes (i) the existence of an established feeder industry, and (ii) that the pastoralists producing the young animals will not compete with them for the available milk supply: if they do the young stock become so poor and stunted that they are unattractive to the feeder industry which must make a profit on its operations.

This also requires a change in the habits of the pastoralists. Developments are now under way in West Africa to sell yearlings from the arid zones to fattening areas in the humid areas along the coast. Heavier stocking is possible only by the use of irrigated lucerne and, or alternatively, silage and hay from fodder crops. With really good management and rotational grazing the carrying capacity can be increased considerably in many areas.

Although most settlement schemes have not been particularly successful, it seems that in Egypt a good deal of sedentarization of nomads and semi-nomads has been successfully accomplished by the Alexandria-Marsa Matrouh project, where these people have been settled and given land, water and seeds. They have been shown how to cultivate their food and plant fruit trees. Special pastures have been set aside for their flocks, and fodder is also grown to tide them over the drought period. High priority must be given to the training of professional and medium-grade range management specialists in each country.

Pasture management

Pasture management involves the improvement of grassland through various practices. It is a most effective and economic way of holding and enriching the soil, provided the pastures are properly developed. It has as yet entered little into African agriculture apart from the improved farming regions on the eastern and central plateau and in the highlands of Africa.

The aim of pasture development should be to raise the yield of animal protein (meat and milk) and other animal products to its optimum economic potential on a sustained yield basis, in such a way that the environment will be protected, including its wildlife and vegetation, and that the best interests of the pastoral peoples themselves will be safeguarded in the development process.

Improved pasture management should also lead to better farming practices and help to prevent soil erosion.

As Africa is the original home of several of the most important tropical grasses which have been widely distributed around the earth and are extensively used in grassland improvement, there are therefore ample species from which to select for the purpose of pastural management. These species should be used in preference to introduced ones, because they are better adapted to the prevailing ecological conditions.

Prospects and potentials

An overall programme of grassland improvement should consist substantially of applying conservation practices and the following lines of management should be adopted:

1. Adjustment of the number and kind of livestock and periods of grazing to the capacity of grazing lands, so that the plants can maintain their growth during the grazing season. De-stocking should be carefully implemented wherever overgrazing exists.
2. Allowing enough growth of desirable pasture plants to cover and protect

the soil and to provide forage for livestock. The deferring or discontinuation of grazing on portions of the grasslands each year during critical periods of growth provides for the production of seeds by the more desirable grasses and for a build up of food reserves in the roots for the next year's growth.
3. Maintenance of a proper distribution of grazing animals so that localised areas are not overgrazed while others are undergrazed. Rotational grazing may relieve trampling and overgrazing in many areas which are normally subject to concentration. Occasionally shifting the cattle by driving or movement between paddocks may be necessary if distribution is not secured otherwise.
4. Increasing the legume population of the grasslands by rest or deferred grazing periods to allow the legumes occurring naturally in the pastures to mature and produce seed or introducing legumes where they have been eliminated. Well-established legumes in grasslands may double or even triple forage production and also improve its nutritive value considerably (CLATWORTHY, 1970). A major problem has been to find suitable legumes for tropical conditions, because there is a scarcity of indigenous species.
5. Conserving and increasing the more desirable woody species for browsing, especially during the long, dry seasons. Drought-tolerant trees make a valuable addition to the available fodder resources during such seasons. Certain trees produce edible pods which can provide an important emergency diet during dry seasons. They also provide shade which is an essential requirement for livestock. The maintenance of an adequate supply of trees requires careful management, otherwise bush-encroachment will result.
6. Correcting deficiencies of major and minor elements in the soil by fertilization programmes. So far this has been practiced little in Africa, but wherever it has been applied it has produced spectacular results. Unfortunately the greater majority of the farmers are either too poor or too prejudiced against this type of improvement.
7. Harvesting surplus forage during wet seasons of maximum growth and storing it to feed to the animals during the dry seasons. Unfortunately it is difficult to get this practice established among most Africans because they are simply not accustomed to it. Presumably this is something that can be overcome with an effort in extension education.
8. Digging wells, drilling boreholes, developing springs and building ponds and reservoirs to reduce the distance which the animals must travel to water, and enabling them to make more uniform use of the areas that are being grazed.
9. Providing complementary seeded or planted pastures of the most productive species and varieties that may be fertilized and irrigated to supply more and better forage during breeding seasons as well as dry seasons and droughts.
10. Developing programmes of planned or prescribed burning. Although research has pretty well established when and where to burn in order to get maximum production, it has proven difficult to put these research results into practice.

11. Training in grassland improvement, plant ecology and range management. More and better training of range managers is most essential. This has been recognized for a long time and some training programmes are now in existence. However, so far, inadequate numbers of range managers have been trained.
12. Pasture improved through careful selection and the use of irrigation and fertilizers could lead to manure becoming more abundant and of better quality, while if fodder crops were grown to supplement dry-season food shortage, a system of mixed farming could be envisaged which could lift the general productivity levels of both agriculture and livestock.

SEMPLE (1971) has expressed the opinion that if grassland conservation practices are put into operation, wherever they are applicable and economically feasible, net production of meat, milk, and other animal products should be at least doubled or tripled.

Animal husbandry

PROBLEMS

Between the different ecological zones livestock vary greatly in number and distribution patterns. Their size and shape range equally considerably between breeds, as for example in cattle from humpless West African dwarf breeds to large-necked humped Afrikanders and medium-sized, thoracic-humped Shuwa zebus. Many European and Indian cattle have been introduced to stimulate milk and meat production and have produced many crossbreeds with local stock. There are many sizes and types of indigenous goats, sheep, pigs, and poultry, of pure exotics and of the various crosses between the two origins (LYALL-WATSON, 1965).

Livestock management varies significantly from extensive nomadic systems, such as those of the Fulani of the Sudanian Zone, through transhumant conditions, as in Masailand, Kenya, to intensively managed peri-urban dairies. Intensive cattle production so far has been possible only in areas unaffected by extreme heat and drought conditions, such as the highlands of East, South and Central Africa.

When discussing productivity of livestock, a clear separation should be made between intensive and extensive production. The maximum production in East-Africa under intensive management, using zebu × Hereford or Aberdeen Angus was 829 kg/ha (STOBBS, 1969). In most areas, and particularly in the arid zones, production is much lower, particularly because of a lack of food and water over extended periods. Under low intensity management the reproductive and growth rate of cattle is slow. Breeding takes place between $2\frac{1}{2}$ and $3\frac{1}{2}$ years and market weight is obtained in 5–7 years. The number of live birth rates rarely exceeds 50 per cent. At least in the underdeveloped parts of Africa beef cattle are generally of poor breeds, maintained in poor condition. It has been estimated that only 30 to 50 per cent of the calves survive each year in some African countries and that only 8 to 10 per cent of the cattle can be slaughtered each year for meat. Carcass weights are 60 to 150 kg, compared with 200 to 250 kg in Europe and America (LOOSLI & OYENUGA, 1963).

Cattle production becomes uneconomic in areas with less than 400 mm rainfall because growth is either checked severely or weight may be lost during the dry seasons.

Much of the drier country in Africa is inhabited by cattle-raising people, like the Masai and Karamoja tribes who live almost entirely on diets of milk and blood derived from their herds. Most of these tribes keep their cattle mainly as symbols of wealth and for the acquisition of wives. The actual number of animals kept tends therefore to be more important then their condition! Cattle are also bred for looks and not for maximum meat production. Ankole cattle from Uganda, for instance, have enormous horns, but are poor meat producers. Although many nomads usually have sufficient or at times super-abundant protein in their diets, not all of them are in this fortunate position. In Somalia, for instance, with 0.6 livestock units per capita of a mainly pastoral population, there are inadequate livestock resources to support the people.

In addition to the reasons already listed for the generally low productivity of livestock in most parts of Africa, mention should be made of some additional constraints, to know the high expense of bush clearance, disease and tsetse fly control, and the difficulties of pasture improvement in low rainfall areas.

Next in importance to beef are mutton-lamb and goat meat. Sheep and goat production is practiced throughout Africa, the latter being a much preferred food in many countries and therefore produced in greater quantities. Sheep are normally maintained by Africans for their meat rather than for their wool.

So far, pork and poultry production has remained relatively insignificant in most countries. Climatic rigours and deficiencies in the quality, balance and amount of food economically available have so far hampered the creation of a large and progressive pig industry. Local breeds of poultry of uncertain ancestry are slow-maturing and poor yielders of eggs and flesh, but they are quite able to fence for themselves. African hides and skins are playing an increasingly important part in the world market.

A great variety of diseases, parasites and insects afflict livestock, presenting one of the greatest physical barriers to improvement. Diseases cause considerable loss both in mortality and in their effect on the condition of the animals. The productivity of livestock is likely to be considerably enhanced if contagious and other diseases are kept under control. Following the disastrous outbreak of rinderpest at the turn of the century, there have been great advances in such control – and efficient vaccines now exist for rinderpest, bovine pleuropneumonia, and other diseases. Tick-borne diseases are being controlled by regular dipping and spraying, which is compulsory in some areas.

Despite the various constraints mentioned, animal husbandry is one of the most promising fields for development in Africa because of the rapidly increasing demand for meat and other animal products. It can be anticipated that with improved production and marketing techniques the livestock industry will utilize most of the available and suitable lands.

Although pasturalists are notoriously resistant to change, in some areas at least their attitudes are gradually favouring a monetary economy, and more sales are being made of their stocks. In the opinion of SEMPLE (1970) the annual offtake under efficient management may be as much as 25 per cent or more of the total number of animals in a herd, without reducing the productive capacity of the livestock enterprise.

Many years may elapse, however, before measures taken to increase the output of livestock products will produce any noticeable effect. Cattle and sheep have a relatively long life cycle and they produce only a limited number of offspring per year. Thus the possible rate of increase in cattle numbers through domestic breeding is not high. Efforts have been made to improve the quality of livestock by introducing more productive breeds or by crossing local breeds with introduced breeds and good results have been obtained with this work, particularly in Kenya. Intensive production is likely to be increasingly based on European type breeds.

Intensification will call for increased and more efficient use of all sources of animal feed. Feedlot fattening and/or fattening on improved pastures could be established. With the addition of fodder and other feed supplements it would seem economical to try out feedlot fattening which should in some measure counteract the loss sustained when animals have to travel over long distances to slaughter houses. As in most fattening areas there is usually a serious competition between man and beast for grain, new fodder crops should be made use of, including oil cakes, fruit pulp and molasses. It seems likely that some day the livestock industry will make use of cottonseed cakes. Silage could be produced to ensure continuous year-round feeding. Feedlot fattening has been tried out successfully in an FAO project in Kenya, demonstrating that this can be done.

The growing demand for cow's milk should lead to improvements in the quality and efficiency of production. As dairy herds in Kenya and in South Africa are highly productive, there is no apparent reason why high yields of milk could not be obtained in other regions.

The control of the tsetse fly is a prerequisite for livestock use of the forests or savanna woodlands; an area of about 10 million square kilometers is still inaccessible to cattle production.

It should be possible, however, to introduce pigmy breeds of cattle, goats, sheep and pigs with a considerable tolerance to trypanosomiasis more widely in tsetse-infested forests or woodlands, to form a valuable addition to family diets.

Where conditions warrant their keeping, goats and to a lesser extent sheep, should be raised for milk and meat. Certainly much better use could be made of milk and cheese of goats and the wool of sheep. Improved breeds should be tested and introduced. Sheep wool yields are generally poor and higher yielding varieties should be imported and cross-bred with local breeds. Higher-grade poultry should be introduced. In order to be economically successful, they must be well fed, watered, housed, and protected from vermin, pests and disease. Under these conditions their production should be far superior to local breeds. The production cycles of pigs and poultry are short, and because they breed prolifically, their rapid expansion is technically possible. Pigs and poultry are efficient converters of various by-products into high quality protein.

The production of a ruminant and its maintenance in good health rests mainly on the value of the ration it ingests. For good production it should be provided with a sufficiently rich and balanced 'menu' the year round. As these conditions are far from being found in most of Africa, too much emphasis cannot be placed upon the need for better livestock nutrition. The possibilities of bringing about improvements in livestock nutrition include the adjustment of animal numbers to carrying capacity of grazing lands and introduction of systems of grazing management, including controlled burning. They also involve improvements in water supplies to ensure adequate utilization of grazing land, and to avoid long journeys to water and a concentration of animals around a few watering points.

It will also be necessary to establish sown pastures to produce fodder in crop rotation, to plant fodder trees and other drought-tolerant species, and to maintain reserves of feed, including high-protein feeds for times of scarcity. Increased production of beef, mutton and goat meat is intimately tied to improved range management and fodder production techniques, and calls for accelerated efforts to secure enclosures and greater security of grazing pressure. Better facilities should be provided, including dipping and spraying facilities to control tick infestation.

Instead of long journeys to slaughtering places, it will eventually be desirable to establish such units near producing or finishing areas and to truck or fly the meat out to consumer areas.

One much overdue change which is now in progress is the blurring of the distinction between pastoralists and cultivators, as more of the former take up some cultivation and more of the latter regard their livestock as an economic asset. Although local integration of efforts between ploughing and manuring the land can be observed, it must be admitted that these efforts are still localized. Mixed farming, in which the manure is used extensively to fertilize the land, has so far not been very successful because most tribal farmers have not been accustomed to this approach. Exceptions to this pattern occur mainly in East Africa: The Chagga people of Tanzania maintain gardens alternated with grassland for livestock to produce manure. In the highlands of Ethiopia abundant use is made of manure. In Kenya and Uganda good use has been made of manure produced by dairy farms in recent decades. Mixed farming should be just as profitable in Africa as it is elsewhere and the increased adoption of this practice undoubtedly holds out considerable prospects for greater productivity. There is need for an increasing degree of collaboration on disease control between countries and for strengthening of national veterinary services. Regular surveys should also be made of the incidence of diseases and parasites among wild ungulates and how diseases are spread back and forth between wild and domesticated herbivores. Despite careful disease control, disease agents will continue to be there and constant vigilance will be required, since we must anticipate surprise attacks at any time.

Future trends

It is clear that in view of the scarcity of protein resources livestock production should be greatly improved and intensified. Technologically speaking there should not be any real bottleneck in the way of progress. Unfortunately, however, it will be a slow process to try to change the customs and attitudes of the nomadic pasturalists in adopting these techniques and therefore it may take decades before livestock production in the arid and semi-arid zones substantially improves. Education and training of animal productionists is therefore urgently needed. In the meantime, rangelands will continue to be overstocked and the downgrading of the land will hasten. Hopefully livestock production on the highlands and in southern Africa will be intensified more rapidly and will be able to meet increasing demands. Considering the prevailing scarcity, however, it seems most likely that protein deficits will increase, rather than decrease.

Livestock developments are bound to have a depressive effect on wild animal populations concentrated on rangelands, as pastoralists will not tolerate competition with wildlife if there is a need to increase their herds.

Forestry

Forest production

The humid atlantic tropics are blessed with favourable conditions for wood production, due to the existence of adaptable fast-growing species, the possibility of production at a low cost, and the proximity of harbours. Although large areas covered by rainforest are still to be found in the Congo basin, Gabon and in the Cameroon, most of it has been cleared or rendered unproductive. Aubréville already in 1947 presented a very depressing picture of the destruction of tropical forest and its replacement by savanna vegetation. Fortunately some forest has regenerated well; however the area covered with original forest, untouched by man's hand, is now quite small.

For example, the high forest reserves of Nigeria cover only two per cent of its total land area (Ogbe, 1966). The area under tropical forest has declined particularly during recent decades because of the encroachment by permanent crops, such as coffee and cocoa, and of clearance for other crops. There is very little hope of increasing the forested area owing principally to the demand for farming both for arable crops and plantation agriculture: the financial yields from cocoa, rubber and oil palm are usually much higher when directly compared with those from timber.

The value of the tropical forest belt is well-known on the export markets of the world. In 1968, for instance, exports of wood and lumber from the Ivory Coast, Ghana and Nigeria were valued at over $ 14 million (FAO, 1969). The richness of this forest has led forest managers in the first half of the twentieth century to concentrate their attention mainly on the economic exploitation of the most valuable species, and on attempts to replace these either by natural regeneration or by various methods of enrichment planting in relatively undisturbed forests.

During recent decades there has been a steady increase of interest in the intensive redevelopment of smaller areas of forest land, and in the utilization of many more species. This is an urgent matter, because the growing stock in these forests is diminishing at an alarming rate. Removal of the more valuable species has resulted in quality deterioration of the forest stands. In order to combat this, improvements have been introduced, including the establishment of management plans, attempts at settling shifting cultivators, and some measure of fire control. Management ranges from sporadic enrichment planting to complete plantation establishment. Enrichment treatments have generally failed to rehabilitate the cut-over forests.

Today's emphasis is on intensive forest plantations. As a result, an entirely different ecosystem is developing, frequently consisting of even-aged stands of a single species. In some countries, considerable new planting or replanting is being undertaken. The advantages of intensive plantations are an increase in the amount of utilisable wood, predictable demands for manpower and money, and concentration of effort in regenerating, managing and harvesting the forest. The problem in reforestation is to establish the young trees at minimum cost, one of the principal expenses being that of land preparation. The utilization of tree species which have so far remained unused, should improve the economics of land preparation. One of the most promising ways of doing this is conversion of the wood into charcoal. Recently, processes have been developed to use most of the tropical tree species for the production of paper. The choice of species to be planted depends on the end product required and on the amount of money that is available. Suitable plantation species include *Araucaria*, *Cedrela odorata*, *Terminalia ivorensis* and *Terminalia superba*. Teak, *Tectona grandis*, is being grown on an increasing scale.

In many parts of tropical and sub-tropical Africa plantations of Australian gums, *Eucalyptus* spp., have been established. These trees grow quickly and may yield good timber; they also facilitate drainage.

Many tropical high altitude areas are reforested with species of pines (*Pinus radiata*, *Pinus merkusii*, etc.) and cedars (mainly *Cypressus lusitanica*), all introduced species. These plantations are highly productive. Disadvantages of monocultures of trees are increased chances of disease outbreaks and damage to young plantations by wildlife and insects.

The dry forests are in many areas important producers of poles and fuel for the local population. Forestry production in the wooded savannas and steppes is hardly sufficient to satisfy even the rudimentary needs of the rural population. This is due to the small volume of standing stock and to the very limited plantations of industrial woods, where there are any. Only a few species in the Miombo woodlands have value as lumber, but much use is made of these woodlands for charcoal production.

Far too little attention is given in current management practices to the utilization of forests on a sustained yield basis, so necessary not only from an ecological point of view, but also for the continued viability of the industries that have been established to exploit and use the resource.

The value of shelterbelts

Shelterbelts, often called windbreaks, are barriers of trees and shrubs planted to reduce wind velocity, evaporation and wind erosion; and to protect crops, homes and livestock. The role of windbreaks in agriculture is increasingly being recognized since many countries have started shelterbelt programmes.

Where trees, shrubs or hedges grow well, in areas affected by erosion, the use of these belts is beneficial both to control erosion and to protect crops from hot, dry winds and from blowing sand or soil.

In addition, they have an ameliorating effect on the microclimate. Specifically, the humidity and temperature of the protected zone increases and moisture may be conserved by reducing evaporation and transpiration. As a result, they usually increase the quality and quantity of the crops which they shelter, such as cotton, tobacco, cassava, groundnuts, and citrus and banana orchards. Shelterbelts serve not only the above mentioned purposes, but also to produce wood for forest industries and for firewood, particularly in those areas which have been largely deforested.

There is a considerable need for the establishment of shelterbelts in the Sudan and Sahel Zones to protect agricultural crops including groundnuts, cotton, wheat and maize. These belts also provide shelter for cattle. In the Guinean Zone, where the precipitation is higher, shelterbelts are nevertheless considered useful.

The most valuable tree species for the Sudanian and Sahelian Zones are *Acacia nilotica*, *A. tortilis*, *A. senegal*, *A. albida*, and *Anogeissus leiocarpus*. Mesquite (*Prosopis chilensis*) has been introduced from North America to the Sudan. This species resists regular cutting, and it grows quickly. The cashew tree (*Anacardium occidentalis*) is used in Senegal as a shelterbelt species.

The planting of fodder trees as a supplement to grazing in arid areas offers promising results in many areas. Species that have been tried for this purpose include the mesquite, the carob (*Ceratonia siliqua*), the honey locust (*Gleditsia triacanthos*) and currajong (*Sterculia diversifolia*).

The use of trees in rejuvenating the soil in dry tropical zones

So far only limited efforts have been made in using trees to rejuvenate the soil. This practice should be resorted to on a much larger scale.

Acacia albida is perhaps the most valuable tree found in dry tropical Africa. It exists in all the African countries with a long dry season from southern Algeria to Transvaal. It has been introduced into West Africa where important stands now occur. Apart from its wood, which is used for firewood and as a raw material for handicrafts, its foliage and pods produce an excellent fodder, and it has the power to rejuvenate the soil. This action, which pedologists have recently brought to light, affects all the physical and chemical characteristics of the soil. This is evidenced by a general increase in fertility to the extent that the harvest of millet grown under *A. albida* can be four or five times greater than grown on the same soil under normal conditions. It is, therefore, of the greatest importance that this species be increased and it should play a major role in the agricultural management programmes of dry tropical zones (Giffard, 1966).

The neem (*Azadirachta indica*), a native tree of India, is another species of great value in arid zones, because it grows quickly, provides shade, fuel, wood for building and furniture, and improves the soil. In India, the bark is much valued for medicinal purposes, and in Ceylon the leaves are used as a mulch for agricultural crops. This species has been introduced into the Guinean, Sudanian, Eastern and Zambesian zones. No doubt more species with comparable qualities will be found and utilized.

Afforestation

Afforestation is one of the best methods of soil conservation in hilly water catchment areas and also an excellent tool in land reclamation, or the rehabilitation of land which has been eroded, gullied, buried by soil depositions from erosion, covered by spoil banks from mining operations, or degraded by bad farming practices.

A distiction should be made in the drier tropical parts of Africa between plantations which have dominant protective functions and plantations which are established for purely productive purposes, such as timber and fuel supply, or for fodder supplement.

Reafforestation is sometimes necessary to prevent the spread of undesirable species. In Ghana, for instance, the guava and *Schweberia mexicana* are invading logged-over stands. An advantage of these species is, however, that they grow rapidly and provide a useful cover of the soil.

Teak is an example of a species used for timber production. It grows well on good soil in the pre-forest savanna zone. The cashew is an example of a multiple purpose tree: it provides good edible fruit rich in vitamins; balsa wood and fire wood.

Future trends and needs

Africa's forests will continue to be subject to increasing pressures from competing forms of land use, for with a growing world demand for forest products, the interest in tropical forests is increasing rapidly.

Future increasing pressures on the remaining forested land will result from: 1. increasing demands for agricultural land, particularly in the tropical parts of the continent; 2. the establishment of new human settlements and the spread of existing ones; and 3. the development of water impoundments, new transportation systems, etc.

The general trend toward a further reduction in forest area will therefore continue on a massive scale and as a result of the increasing demand for land for agriculture and grazing, very little lowland tropical forest will be allowed to remain.

Increasing use will therefore have to be made of the timber resources. It will be necessary at the same time for efforts to be made to control exploitation in such a way that the forests will eventually be renewed. For this reason there is an increasing need for intensively managed plantations of one or a few species.

Special measures should be taken to ensure that the introduction of intensive forest management techniques and the extension of manmade forests will not cause the disappearance of useful forest genetic resources and of forest wildlife, or seriously impair soil and water values. More attention should also be given to future recreational and aesthetic needs for forestry.

Natural protection forests and protective plantations must be managed with the objective in mind of improving water yields, stabilizing soil and conserving its properties, and preventing erosion and desiccation. These protective forests should be maintained permanently and the harvesting of the timber and fuel must be only a secondary consideration.

The following practives should be undertaken: 1. Bringing more forests into use and production. This means on the one hand better management and use of the closed rainforests and on the other hand establishing new forest plantations in the savanna areas. 2. More should be produced on the forest land already in use. This means improving existing stands and creating more quick-growing plantations as part of an already productive forest area. 3. Better use should be made of forest products, including the establishment of modern integrated wood-using industries and using more fully all removals from the forests by employing the latest techniques. More ways should be found of using economically a larger number of species.

Forest land tenure should be developed to arrive at a balanced pattern of ownership. Government ownership has special importance in implementing long range policy, since fragmentation is not as likely to happen as under private or tribal ownership.

One of the most serious obstacles to development of forestry and forest industries in Africa is the shortage of trained staff at all levels. For this reason training programmes should be improved and enlarged with international and bilateral technical assistance. FAO has been and continues to be active along those lines.

Increasingly more long-term plans are being made for heavy forestry investments without the benefit of a master plan that would safely co-ordinate planned forest land use with plans for meeting the country in future resource needs. This should not be allowed to continue, because properly conceived land use plans are needed to help insure that the trees the forester plants today can be expected to grow to maturity and provide the raw material needed by industry in the future.

Inland fisheries

Problems

One of the main problems with the development of inland fisheries has been the lack in adequate recognition of the importance of this resource. The economic values of Africa's potential for fish production are only slowly being realized and although certain inland fisheries have been fairly well exploited, many streams and lakes remain very much under utilized.

Almost all natural bodies of water and most artificial ones support fish life, and when carefully managed, fisheries resources are a valuable renewable asset. In tropical Africa the total production of inland fish almost equals that of marine fish, an average yield of 2,000 kg/ha per annum seems easily attainable. Fish production from inland sources in Africa in 1969 reached at least one million tons, and represented at least $ 100 million, which was circulated entirely within the African economy (FAO, 1968).

Improved methods of fishing in rivers, lakes and reservoirs, plus the capture of currently unutilized species are likely to lead to an appreciable increase in landings. In addition, several successful management measures have been devised to substantially increase production. These include mainly better catching techniques. The maximum use of inland fishery resources and culture of fishes and shellfishes assumes particular significance in a continent with a chronic lack of protein. Their most important role in the developing countries of Africa is undoubtedly provision for domestic consumption.

Fish culture in fresh and brackish water areas can contribute substantially

---

Photo 28. About 90 percent of Dahomey's population lives in villages. The Government's multipurpose development plan is aimed at raising the standard of village life. Fisherman working on an FAO fisheries project on Lake Ahémé. Photo: WFP/FAO.

towards meeting protein deficiencies of people in rural areas besides providing employment. If fish were produced in ponds or similar water areas throughout a country, fresh fish would then be available even in remote areas.

There are all over Africa extensive areas such as freshwater and saline swamps, spill areas of rivers, natural depressions, which are suitable for pond construction and experience in many countries has shown that one of the most profitable ways of using swamps is to convert them into fish ponds. Rice fields, irrigation and multi-purpose reservoirs, as well as irrigation and drainage channels can also be used for fish culture. Commercial fish culture can be undertaken by individual cooperatives or the government. Fish culture in rice fields not only produces a fish crop, but because of the excrements of the fishes also increases yields of rice. Small ponds constructed for stock watering or irrigation can often also be used for fish production.

A completely satisfactory pond fish has so far not been established in Africa and detailed studies are necessary to select suitable species. Experience with *Tilapia* and certain strains of common carp has not been entirely satisfactory, although those strains that breed more than once in a year may offer possibilities. The culture of larvicidal and snail-feeding fishes has also a great importance from the point of view of public health under African conditions, because of reduction in the incidence of malaria and bilharzia.

There are possibilities for substantially increased production from coastal aquaculture. So far, work in this field has in most countries not progressed beyond the experimental stage, although the commercial feasibility of oyster cultivation has been demonstrated in Senegal. The existing production of cultured fish and shellfish from coastal waters remains insignificant as compared with its potential. Huge areas of saline swamps exist which could be converted into productive fish ponds and it is estimated that in the Niger delta alone 750.000 ha could be used for that purpose.

Man induced changes in the aquatic ecosystem have had a marked impact on the fish supply. These include: 1. the change from a river-flowage system to a lake-gathering ecosystem in hydroelectric dam projects, creating large manmade lakes; 2. the pollution of rivers or lakes by industrial discharges, commercial fertilizers and pesticides, drilling and mining products; 3. the change of habitats by drainage of estuarine marshes; and 4. the massive growth of aquatic plants associated with sedimentation.

The insidious and cumulative effect of various environmental changes on fish production could be serious. Affected also could be spawning areas and the movements of migratory fish. Fortunately, techniques are available to combat some of these ill-effects. These include the installation of fish facilities in dams, fish diversion devices at the heads of irrigation systems and protective devices in power plants, together with measures for environmental improvement in upstream as well as downstream areas of dams.

River basin development projects, including the construction of dams for water shortage and power development or diversion of water for irrigation, have very seriously affected inland fishery development. Besides being a

Photo 29. Installing nets on Lake Ahémé as part of a FAO fisheries project. Fishermen receive World Food Programme rations of dried fish until such time as the cultivated fish mature in the lake.

mechanical barrier to migration essential to the life cycle of certain species of fish, a dam can cause hazards to fish life due to the changes brought about in the ecological conditions such as reduction of flow, alteration of temperature regimes, and changes in the chemical properties of water in the downstream areas of the dam. In the upstream areas deep inundation of spawning grounds and shallow water habitats occur, as well as violent fluctuations in water levels affecting survival of young fish and spawn, bottom inhabiting animals and aquatic plants needed by fish. On the other hand, several species of fish flourish quite well in large reservoirs created above dams, and can provide productive fisheries. Increases in inland production have been achieved also through more intensified fishing of some of the larger African lakes.

Trends, needs and potentials

Africa has many interior wetlands and one of the most promising lines of development has been improved management and enlarged use of their fish resources. Zambia is an excellent example of achievement in improved standards of nutrition for urban populations from the freshwater fisheries of the numerous swamps.

With a view to increasing the area of land for agriculture or for urban expansion, swamps, lagoons and deltaic areas are being reclaimed on an increasingly large scale. The draining of wetland areas will eliminate valuable nursery grounds and the channeling of river courses for drainage purposes will have a highly detrimental effect on fish life.

The growing danger of pollution of inland waters is a problem of the utmost importance in the maintenance and development of inland fishery resources. With industrialization taking place and use of fertilizers and pesticides in agriculture increasing, this is becoming more serious as time goes on. Direct changes in freshwater environments have resulted from deforestation and over grazing, bringing with them floods, erosion, and heavy siltation. Diversion of water for irrigation purposes has reduced fish populations. As irrigation will assume much larger dimensions in the future, this will no doubt have an increasingly negative effect on fish production.

In many areas experiments continue to determine the best stocking practices of fish ponds and the value of fertilization. The creation of more fish ponds can well be considered as one of the best and cheapest measures of increasing animal protein.

The presence of large lakes and marshes in East Africa has precipitated the development of a large fishing industry. In West and Central Africa substantial fisheries exist on the Niger, in Senegal and in Zaïre. There are many indications that an important increase in production is possible, estimated at 600,000 tons per year in Africa south of the Sahara. Lakes Tanganyika, Rudolf and Chad are under-exploited and a better use of their fish populations would add considerably to the total fish production of the continent.

Efforts should be made to industrialize fisheries. More outboard motors should be provided for canoes and new boats with good engines should be constructed. A special effort should be made to improve methods of treatment and preservation of fish including drying, smoking, etc. Cooperatives should be set up as individual fishermen can achieve very little.

Although research and demonstration projects on the pisciculture of *Tilapia*, *Heterotis*, carp and blackbass exist, this type of research could be intensified with advantage.

The fisheries of the Great Lakes of East Africa, a special problem

The Great Lakes of East Africa, Lakes Victoria, Malawi, and Tanganyika, are the repositories of the world's three richest lacustrine fish faunas. Nowhere else are there lakes that even begin to rival these in their diversity of fishes. They are the sites of fisheries producing many thousands of tons of fish per annum in a region where this is a much needed commodity. Unfortunately, this fabulous resource is in trouble already: The fishery of at least one major lake, Victoria, is being inefficiently and dangerously exploited. In this lake an important fishery is based on the catch of cichlid fishes belonging to the genus *Tilapia*. Already the rewards from fishing have been dramatically reduced from a catch of about 25 fishes per net in the Kavirondo Gulf at the beginning of the century to one below 0,5 fish per net in some areas – and the individual fishes caught are now much smaller than formerly (FRYER, 1972). The available facts are clearly indicative of gross mismanagement.

Another man-induced problem in Lake Victoria is the introduction of some fish species, among which the Nile Perch. The addition of this predator, who preys on herbivorous fishes such as *Tilapia*, reduces the efficiency of the food-chain productivity considerably and its introduction can be considered a tragedy (FRYER, *op. cit.*).

Wildlife

PROBLEMS

Although Africa is still the most productive continent for wildlife from a species and numbers point of view, there are vast areas from which many species have been eliminated or where wildlife has been reduced to such a low level that it no longer plays an important role as a resource.

The need to protect wildlife for aesthetic reasons is well recognized. During recent decades, however, African wildlife has been increasingly managed for economic reasons associated with benefits to be derived from the tourist industry and protein production.

As human populations expand, there is an increasingly serious risk that, even in such havens as East Africa, wildlife will be severely reduced or will be largely confined only to designated 'wildlife' areas, such as national parks. Considering the aesthetic importance and tremendous economic value of wildlife to tourism, potentially this could become nothing less than a tragedy, because most wildlife reserves are not viable ecological entities, and often wild animals are killed when they leave these areas for a greater or lesser period.

The availability of a wide spectrum of wildlife species, each using a slightly different niche in the ecosystem, provides the opportunity for diversifying animal production and hence achieving more efficient use of the various available habitats. Although competition for food and space occurs between wild and domestic herbivores, skillful management can increase overall production when a mixture of domestic and wild herbivores is stocked because both groups have different food preferences. Some species of wild animals such as the eland, addax and oryx are less dependent upon water than domestic stock (TAYLOR, 1969) or are more tolerant of such diseases as *Trypanosomiasis* – the scourge of cattle – and can therefore be used in areas that are unsuited to stock or where development for ranching purposes is expensive. The argument that wild herbivores cannot be maintained on rangeland used for livestock is unrealistic, because there are many examples to show that certain species of wildlife can be effectively produced in combination with

Photo 32. A zebra in East Africa. Some wildlife species can under certain conditions yield more food than domestic stock. Game has a greater rate of population growth and can thus be harvested more frequently than domestic stock. Photo: FAO.

domestic stock, DASMANN (1964), for instance, states that many ranchers in the Transvaal have reduced the number of sheep or given up sheep raising altogether in favour of raising game for the market.

Many legislators and administrators find it rather difficult to believe that something as commonplace as wildlife, with no specific owner, can have an intrinsic value capable of exploitation for the good of the community, or that its disappearance is a real loss. Perhaps this is one reason why nature conservation agencies, even in countries with potentially valuable wildlife resources, are often given a low priority among government organizations. This

---

◁ *Photo 30.* Botswana is richly endowed with wildlife, and this resource still contributes considerably to the national economy. Female waterbuck and lambs in the Chobe National Park. Photo: FAO.

◁ *Photo 31.* The future place of wildlife on Kenya's rangeland has been carefully studied by UNDP/FAO projects. Animals are sometimes hunted from a helicopter and stunned with a drug-filled syringe shot from a gas-powered pistol, so that they can be examined and marked for future recording of their movements. A wildlife biologist is taking a tranquilized zebra's temperature. Interested Masai warriors look on. Photo: Kenya Ministry of Information, issued by FAO.

Photo 33. Commercial advantages from controlled game cropping include the trade in wild animals and the sale of hides and skins, and similar animal products for export. Man examining a crocodile skin ready for export. Photo: FAO.

viewpoint of administrators also helps to explain the tendency to sacrifice nature conservation principles rather readily to political or administrative expediency (Child, 1971).

If any progress is to be made in wildlife management there is a need for more general acceptance of controlled wildlife utilization as a legitimate form of land use in the rural economy. Such an economy is ultimately dependent on the ecological stability of the range and on the wide spectrum of wild animals and plants which has evolved in the arid zones. In many of these areas, wildlife utilization in its various forms appears to offer opportunities for diversifying land use cheaply, without further endangering conservation values (Child, 1970). The rationale behind wild-game cropping as another use of the land has been well summarized by a number of authors (cf. Talbot, 1966 & de Vos, 1973). Spotted cats, like leopard, cheetah and serval (*Felis serval*), have also been killed in large numbers and sold for their valuable furs. As a result, the cheetah in particular has diminished in number greatly. For this reason, the killing pressure on these cats should be reduced and neighbouring states should coöperate in an effort to stop illegal export of poached skins.

Trends, needs and potentials

Game laws and policies need revision in order to enable governments both to nurture commercial activities through legalizing the sale of game meat and wildlife by-products, and to protect the resource from depletion. Neighbouring countries should make more efforts to align their laws so that illegal import and export problems, and poaching can be minimized along international boundaries.

Game ranching projects should be developed in those parts of Africa where they are not now existing to study the possibility of an alternative to grazing cattle and other domesticated animals or a combination of livestock and game ranching in suitable areas. Game ranching appears to be particularly profitable in areas which are too dry for cattle production or in tsetse infested areas.

A range of highly productive species of ungulates adapted to native conditions can be made to yield without difficulty acceptable protein and other useful products on a sustained yield basis. The value of wildlife is especially important as one means of raising production in areas that are unsuited to intensive agriculture, or where other more conventional forms of land use are impractical or endanger essential conservation values. In arid and semi-arid marginal and sub-marginal lands, wildlife tends to be less destructive to habitats than domestic stock. This is because it has evolved under local conditions and is therefore more adapted to the habitat than introduced stock. In many instances the utilization of wildlife requires relatively little development capital when compared with other forms of land use, and the demand for its products has been amply demonstrated in several countries, such as the Republic of South Africa, Rhodesia, Zambia and Kenya.

It should be emphasized that not all wild species will be usable in the context of game ranching. Some species such as buffaloes may be incompatible with cattle. However, I believe that the eland, the oryx (*Oryx beisa*), impala (*Aepyceros melampus*), Grant's and Thomson's gazelles have considerable ranching potential. There is now in Kenya a UNDP/FAO project in the process of demonstrating this as well as the economic value of wildlife versus that of livestock. Efforts to domesticate eland have proven successful (POSSELT, 1963) and it is believed that this species should be used widely for game farming in arid zones.

Reference should also be made here to the value of wildlife as a tourist attraction and consequently as a source of foreign exchange. This value is continuously and rapidly increasing, and in East Africa in particular this has become an important consideration. In Kenya, for instance, in 1971 more than $ 67,000,000 was earned in the tourist industry, which exceeded the returns obtained from export of agricultural products, including coffee and tea.

How can planning for wildlife resources be fitted into the total picture of land use planning? This can only be done after a survey of the capability of land for the production of various other resources has been completed, and a coordinated picture of the possible combined uses has been arrived at. On the less productive land, wildlife should be considered and maintained as an important resource and specific plans should be developed for its management and use. This is not to say that wildlife should be neglected on the better lands, but that the other resources should be given priority. If there are insufficient national parks or game reserves, new areas should be so designated.

There is unquestionably a trend towards more intensive use of rangelands, and it seems obvious that some species of wildlife will have to be eliminated in the process. For instance, it seems that large carnivores, such as lions, and also some large ungulates, such as elephants, will have to go from all areas not designated as wildlife reserves. Several species of large ungulates like buffaloes and zebras which can seriously damage fences may also have to be locally controlled. Other ungulates that are hosts to diseases affecting livestock may have to be locally eliminated.

Soil and water conservation

SOIL CONSERVATION AND EROSION CONTROL

Since the economy of any country depends primarely on the soil and its products, and thus on its crop and livestock producers, soil erosion and depletion will seriously depress the development and the progress of any nation. It is therefore obvious that adequate erosion control practices be followed if a satisfactory level of production is going to be maintained.

Beginnings have been made toward the control of soil erosion in many

African countries, but on many uplands where overpopulation and cultivation of steep slopes have already gone far beyond the point of safety, only scientifically designed terracing and contouring, planned reforestation, careful protection of vegetative cover and management and control of rain and stream water can save them from further destruction.

Terraces on the contour, sustained by contour cultivation, have been established in several countries, as, for example, Tunesia and Kenya. They provide effective obstructions to the flow of water and serve to decelerate erosion, but as they are not permanent features they do little for soil improvement.

The surest and simplest way to maintain the soil in a productive state is simply to provide and to perpetuate cover, which can often be accomplished even on poor and droughty soils.

Cover crops which add organic matter and improve soil productivity, like lucerne, other legumes and selected grasses, need to be planted or seeded. Most grass crops are good soil builders, particularly those which are allowed to grow for several years. Cover crops are surprisingly effective in controlling soil blowing. Although many African tribes did use these crops widely under primitive agricultural systems, they were later discouraged from doing so by Europeans who introduced cash crop farming. It will now require considerable extension-education effort to re-establish or introduce the widespread use of cover crops.

Tillage should be at a minimum, that is, stirring the soil as little as possible and leaving most of the crop residues on the surface, which are plowed under to maintain organic matter, soil structure and aeration. Crop rotations, when properly used, are also strong factors in erosion reduction. In general, the longer the land is in sod, the greater is the improvement in soil structure. Drainage areas should be in grass, strip cropped or terraced; if not, soil deposits will fill the drainage channels.

Drifting sand and sand dunes are very difficult to stabilize, but on very deep sand, accumulations of mulches of straw, hay, brush, or other vegetative matter can be used effectively as an anchor. The mulch should be applied on the windward side first and locally adapted grasses, shrub or trees should be used for permanent stabilization. Livestock grazing must be diminished if dune fixation is contemplated.

So far only few of the soil conservation and erosion control measures have been applied in those parts of Africa which are in most urgent need of such improvements. The application of oil mulches, which has the added advantage of reducing evaporation losses, has been quite successful in Lybia.

Considerable progress has been made in the Republic of South Africa with the application of soil conservation practices: Terraces, internal farm fences, conservation and irrigation dams, stock-watering systems and anti-erosion structures are all being applied successfully. Detailed farm management plans have been drawn up for individual properties to ensure that control practices receive the essential support of scientific conservation farming methods.

Photo 34. Exposed root system of a tree as a result of erosion by the Luangwa river, Zambia. Photo: A. DE VOS.

In Tunisia and Morocco considerable United Nations Development Programme assistance has been provided in recent years in efforts to improve conditions on badly eroded lands (see photo's 35, 38). This should set an example which could be followed by other countries in which similar arid conditions exist.

The control of wind erosion

The first requirement necessary for the control of wind erosion is the reduction of the speed of wind near the soil. Soil must be held in place and this can best be done by means of vegetative cover. Experimental work in the Sahelian and Sudanian zones indicates that living screens of trees and/or shrubs are effective wind breaks, offering resistance to moving airstreams. These are arranged in lines perpendicular to the direction of the dominant winds, but as wind gradually builds up to its original speed once it has passed a windbreak, there is a need for a network of parallel windbreaks. Their action is two-fold: not only do they reduce the wind speed, but a modification of the micro-climate is brought about with consequent improvements in plant production (see also p. 179).

There is a considerable need for large-scale development of windbreaks or shelterbelts in many parts of Africa.

Soils can be and often are, degraded in productive value. However, they can also be improved by skilled management and the application of the required fertilizers. Unfortunately, thus far not much effort has been made towards soil improvement in Africa.

Environmental conditions in most of Africa are generally not conducive to the maintenance of a high fertility level of the soil, because it is usually subjected to either high temperatures or heavy precipitation, or a combination of both. High temperatures during the dry season followed by heavy precipitation during the rainy season have a leaching action on the topsoil and cause rapid chemical and bacterial decomposition of minerals and organic matter.

This leaching is especially serious when the soils are not covered with vegetation. Yields decline and structure deteriorates rapidly under cultivation.

The management of the old residual soils of the tropics for sustained high levels of production of annual food crops of good nutritional quality is difficult to achieve. These soils are generally lateritic, typically deficient in plant nutrient elements and high in aluminium and iron. Effective improvement of such soils on an economically justified basis is virtually out of the question.

Leaching and depletion of soils, particularly in the tropics, might be reduced through controlling the intensity of agricultural use; by providing fallow periods; by introducing horticulture or pasturage instead of short-cycle crops; or by the use of appropriate fertilizers. Particular attention should be given in erosion control programmes to reducing sediment delivery to streams, reservoirs, dams and other water bodies in order to control physical, chemical and biological effects of these sediments on water quality and aquatic resources.

Green manures and mulches

The practice of adding fresh, undecomposed green plant material to the soil is known as green manuring. In some cases the plant is grown and dug in or ploughed under where grown; in others it is cut and put on to other areas as a mulch.

The primary benefits of a mulch lie in its ability to reduce soil water losses, keep down soil temperatures and control erosion rather than in the addition of nutrients. Mulches are therefore particularly valuable when and where water supplies are marginal.

While it grows, the green manure crop protects the soil and functions as a cover crop. When green manures are added to the soil they are thought to supply organic carbon, nitrogen and a range of other plant nutrients and to have a stimulating effect on soil bacterial activity. Although the theoretical possibilities and benefits of green manuring in the tropics are considerable, inadequate information is available about its best use in practice or to what extent the tropical leguminous cover crops and green manures now used really add extra nitrogen to the system. Too little careful experimental work has as yet been carried out in West Africa on which generalizations might be based as to the extent to which additional nitrogen can be added to the soil by this means, or the extent to which green manures and mulches raise organic matter levels in the soil more than very temporarily.

Reclamation of eroded and abandoned land

Reclamation of 'lost' land will never be easy, rapid or cheap, and indeed, some of the worst of the maltreated land can not possibly justify the exertion or the funds required for its rehabilitation.

However, several countries have found that in conservation work which is justifiable there is much to be gained by fostering closer cooperative relationships, especially between neighbouring countries sharing similar problems of, for example, erosion by water, wind, or both.

There is really little excuse for the neglect of soil resources, because land

Photo 35. To fight soil erosion, trees are planted in Tunisia along the contour lines of the terrain. Cypress, eucalyptus and Aleppo pines are used; they also act as windbreaks and sometimes as property boundaries.

Photo 36. Spineless cacti have been planted as an anti-erosion practice on abutments as an extra reinforcement. These plants keep water from running downhill, sweeping good soil away. Spineless cacti can be fed to cattle and sheep during dry periods. Photo: FAO.

use techniques, combined with engineering and management practices, are thoroughly developed and widely known. Contour practices, designed to fit the topography, combined with soil-saving rotation planting and proper fertilization, provide protection to the soil, conserve water for plant growth, and raise yields of cultivated crops. These are basic practices which can be adopted inexpensively, and in a reasonably short time, wherever they are brought to the attention of even primitive agriculturists. These practices, with others which follow naturally as the principles of soil conservation become understood, are especially adapted to lands subject to erosion by water. They slow up the flow of water, resulting in benefits both easily demonstrable and easily understood. They are particularly well adapted to the reclamation of eroded or abandoned land.

Development of water resources

Man's efforts have in the past been largely directed at utilizing surface and ground water resources. More attention should be paid to the fate of rain water after it has fallen, because its availability to agriculture is to a consider-

Photo 37. Tunisia is affected by a rapid, ruthless erosion which is turning thousands of acres of once fertile land into a desert. To stop this process, which in the long run could reduce local populations to famine, Tunisian agronomists are now trying various types of culture. Seeds of grass are sown in a field to stop erosion of the soil. Photo: FAO.

able extent dependent on man's use or misuse of the land. It is at this stage that control of water, or more particularly the lack of it, makes its most spectacular and extensive impact on the vegetative environment. Through improvements in the use of rain man is able to manipulate the resources of surface and ground water better to provide more domestic water supplies for both himself and his stock, irrigation for his crops and power for his industries, besides other benefits including fishing, navigation and recreation.

One of the most pressing issues in agricultural development is the need to provide adequate supplies of water to farming communities for domestic use, livestock and crop irrigation.

Irrigation developments

Irrigation can supplement rainfall in those areas with long dry periods, and increase the efficiency of livestock production considerably. The development of irrigation is the most important single factor in raising yields and reducing the hazards of livestock production. Irrigated crops can produce very high yields, but this requires careful management. Much controversial thinking

Photo 39. Much of Tunisia is affected by rapid erosion which is turning thousands of acres of once fertile land into a desert. To stop this process agronomists are now trying various types of culture which could eventually be extended to entire arid and semi-arid regions, feed the local populations, fix the soil and stop the deadly progress of the desert. Plantations of eucalyptus and fruit-trees on irrigated soil. Photo: FAO.

still exists on the subject of making water available for crop production. Traditionally the majority of farmers grow only one rain fed crop. Yet, small but significant changes have taken place in their attitudes in recent years and the notion that dry season farming on small irrigated plots can be profitable appears to be gaining acceptance. Double cropping of rice is becoming an accepted practice in certain areas, notably in the northern Ivory Coast.

Water control schemes for irrigation are being developed along the Niger and many other rivers and have been well developed in North and South Africa. Present trends indicate a steady increase in irrigation development in the semi-arid savannas, particularly those of West Africa but excessive use of underground water should be prevented as much as possible. In the Republic

---

Photo 38. In the lower ranges of the Rif Mountain of Morocco trees are planted on bench terraces to help consolidate the soil and prevent its being carried away by rains. Multiple cultures include fields and olive tree plantations protected by bench terraces. On the hills are remnants of the natural forest vegetation. Photo: FAO.

Photo 40. One of the many windmills which have been installed in the United Arab Republic to pump sweet water from newly built wells. Brush stakes in the dunes control the movement of sand by the wind. Photo: WFP/FAO.

Photo 41. Irrigation in the Rhurb plain, Morocco, is expected to raise the people's income considerably. Agricultural improvement includes expanding eucalyptus and pine forests in order to develop timber and paper industries, acclimatizing new fruit trees, expanding citrus production, introducing sugar cane and new cereal varieties, and improving animal husbandry. A new forest of four-year old poplars. Photo: FAO.

of South Africa, where extensive irrigation schemes exist, water is already being drawn from subterranean sources more rapidly than it is being replenished, which is of course a source of serious concern.

In large-scale irrigation projects the improvement or maintenance of soil fertility over a long period raises many problems. As these projects are often in arid areas where a high evaporation rate has kept salts close to the soil surface, irrigation tends to redissolve these salts and deposit them again, resulting in increasing salinization of the arable layer. This may be followed by large-scale land desertion. Both salinization and water contamination are

Photo 42. Fruit trees being irrigated at Ouled M'Harnad experimentation centre, Tunisia. Photo: FAO.

frequently associated with drainage difficulties and to keep irrigated soils free of toxic concentrations of salts, sufficient water must be applied to flush out these salts.

Management practices for the control of salinity and alkalinity include, in addition to making the right quantities of water available, choosing crop varieties which will provide satisfactory yields under moderately saline conditions; the use of land preparation and drainage-tillage methods to help in the control of removal of soluble salts or alkali; special planting procedures to minimize salt accumulations around the seed; selecting the proper irrigation methods to maintain a relatively high soil-moisture level and at the same time allowing for periodic leaching of the soils; and the installation and maintenance of water conveyance and drainage systems.

Irrigation, whether supplementary or year-round, will have to be relied upon in increasing measure to provide sufficient food for the hungry people. It is a prime example of how water control can affect man's entire environment, for the productivity of the land can be increased enormously, bringing about closely knit, prosperous communities.

Much progress has been made in recent years, particularly in Israël (Boyko, 1968) in developing food and fodder crops – including vegetables such as tomatoes and beets – which can be produced economically on soil with low to medium salt concentrations. In the more distant future, improvement of irrigation methods with brackish water should allow for the utilization of wells which are at present useless.

As regards irrigation developments, failures have frequently been high. Inadequate attention was often paid to the conditions which predispose the transmission of waterborne diseases and the breeding of noxious insects. In East Africa scores of water control structures have been abandoned or even wilfully destroyed, since little or no attention was paid to the development of the lands they were meant to irrigate or because they did not satisfy the needs or wishes of the local peasants and their herds.

In the Republic of South Africa many dams designed to provide water for irrigation have silted up and inadequate alternate sites are available for the construction of other dams.

On the continent as a whole the development of ground water must continue to progress if the needs of men and livestock are to be met. The demand for ground water will increase rapidly, and the development of a market economy will allow a larger utilization of underground water for irrigation pumped from tube wells.

For the immediate future, small and medium surface schemes should be encouraged. An interesting development is the construction of hafirs or small impoundments for water storage for livestock. Their size has been set only after determining the forage available in the surrounding area and water is provided only for the limited number of animal days use that the vegetation can safely sustain. Thus, when the water in a hafir is exhausted, the people are forced to shift to another area.

In the longer run, larger projects for flood control and irrigation should progressively assume a more important part in the development of Africa, especially in the tropical areas which are well-watered during part of the year but where the dry season is extremely marked.

Water utilization problems

Most of Africa is already in critical need of improved water supplies and this problem will become much more acute with the rapid increase of human populations and the increasing trend of people to aggregate in urban communities.

This critical need of water as the most vital and limiting living resource in the African environment has been inadequately recognized by those making decisions regarding resource management problems. However, a recent United Nations report (1973) states that there is hardly any area in the continent where ground water cannot be found at greater or lesser depths. Most of this groundwater is deemed acceptable for human consumption and for livestock watering, but it is not known for how long such water supplies can be utilized before they are exhausted.

Photo 43. The Niger is the great river of West Africa with a total length of about 4000 km. At Bamako (Mali) it has a depth of 2 m with a breadth of 435 m. The conglomeration of people washing themselves and their clothes and the watering of animals, along the banks of the river, is leading to increasing levels of pollution which may endanger the valuable fish resources. This photo was taken in Niger. Photo: FAO.

The Food and Agricultural Organization report on 'The Influence of Man on the Hydrologic Cycle' (FAO, 1969) cites a number of common instances of land misuses, pointing out that most water scarcity problems stem from a lack of knowledge of and experience in sound soil management practices. The pattern of land deterioration in arid environments and its effect on the availability of water is clear: excessive grazing pressure reduces the ability of the soil to accept rainfall and encourages run off and water loss. This results in a poorer growth of grass, causing further overgrazing and increased trampling by stock.

The problem of rainfall acceptance by the soil is, of course, not limited to grazed areas. High rainfall areas, generally covered by forest, are of great major importance in maintaining stream flow for less favoured areas downstream, where permanence of the flow is essential for the continuation of life or for the supply of water to irrigation projects. The loss of water in surface run off during rainy seasons as a result of deforestation leaves less water in temporary soil storage which can be released slowly as dry weather flow.

Mismanagement of the land, then, has a direct effect on the availability of water. But it also affects the influence of water on the land by accelerating erosion. It is a fact that water erosion affects much of Africa in spite of the continent's extensive arid areas. This is because so much of the usable land is without protective vegetation for part or all of the year, or has only sparse cover at any time. The most widely prevalent form of erosion is sheet erosion, but in places of even moderate rainfall there is also extensive gullying. In some places tremendous amounts of sediment are removed from top soils by surface runoff.

The cumulative effects of misuse of the land, and consequent erosion and sedimentation in streams and lakes, have affected the quantity of water available for use. Another problem is that the quality of water is also greatly influenced by man through the use of fertilizers, herbicides, pesticides and other chemicals that do not readily break down into innocuous products. In the less developed parts of Africa, which unfortunately cover the greatest portion of the continent, sewage is usually being disposed of in pit latrines, by dumping of night soil into a river or sea, or by other primitive means. This, of course, implies the danger of the spread of water-borne diseases, such as cholera.

As a result of heavy use, the groundwater level has dropped and as it continues to recede many wells have dried out or it became uneconomical to remove the water. In addition, at greater depths saline water was often encountered. The main mistake that is being made and causing rapid depletion of ground water is that the water requirements of crops are almost always persistently underrated. Continued deep-well pumping in South Africa has caused such an increase in the salt content of water that it had to be brought to a complete stop.

The discovery of important quantities of underground water in the lower Algerian Sahara has led to an abusive exploitation of this water with a disastrous effect on the superficial water table. Because this part of the Sahara is shaped like a saucer, the subterranean outflow of water is not assured. Hence waterlogging takes place, causing the death of many palm trees (ACHI, 1972).

Development plans of poor countries must take water into consideration, particularly if it is a limiting resource. The plans elaborated for water development are not only lagging behind the population growth, but most of them are inadequate from the start and do not recognize the true magnitude of the needs.

In a continent which is scarcely endowed with water resources, it stands to reason that efforts should have been intensified to make optimum use of all available freshwater resources. For this reason, many man-made lakes have been constructed, some of which are well known because of their enormous size: lakes such as Aswan, Kariba and Volta.

Many mistakes have been made in river basin development in Africa, largely because these programmes have tended to concentrate on reservoir and irrigation projects and have paid lip service to ecological considerations.

Generally, the primary purpose for which dams are constructed is to provide hydropower, irrigation, or flood control. Broad environmental consequences such as the impact on fisheries and on recreational facilities are usually inadequately considered. The problem that usually exists is that in an analysis of the costs and benefits of such projects too little attention is given to secondary benefits or to the impact which the development of primary ob-

Photo 44. Construction of the Volta River dam in Ghana has created what is claimed to be the world's largest man-made lake where once some 85000 people farmed. Hydro-electric power obtained from the dam is of vital importance to Ghana's industrialization but the resettlement of 12000 farming families has presented the government with many problems. View of the spillway of the Volta dam and part of the huge lake built up behind it. Photo: FAO.

Photo 45. A work boat of the Volta River Authority, Ghana, visits an island (once a hilltop) in the lake created by the Volta dam. The maize crop suggests that a considerable area of productive river bottom land has been flooded. Photo: FAO.

jectives has on the existing natural resources. The result is that unnecessary damage is done to those resources. Take for instance the effect on the fish resource. Not only do dams block the movements of fish up and down stream, but they are often destroyed in turbines below the dams. Dams usually also alter the water regime downstream, which may also adversely affect fish populations and sediment deposited in the reservoir may result in a considerable reduction of food for fish below the dam. Reductions in downstream discharge can disrupt feeding and spawning grounds. Thus, alteration in the downstream water regime may have a big impact on existing fisheries (see also p. 184).

In the case of the Aswan dam in Egypt, deposition of silt in Lake Nasser has apparently resulted in a sharp decline of a sardine fishery in the eastern Mediterranean sea and a reduction of agricultural production. Year-round irrigation has produced not only favourable conditions for agricultural production, but also ideal conditions for the snail that carries bilharzia! This has spread more extensively in a country already badly affected by this dreadful disease. It is clear therefore that the dam is a mixed blessing.

It can be argued that a decline in downstream fish resources can be offset by the development of a successful fishery in the reservoir. However, because of the general inability of riverine fishes to adapt themselves to the new conditions created in man-made lakes, the introduction of other, better adapted species is usually necessary at considerable cost. The often severe drawdown in reservoirs can destroy spawning grounds and feeding areas of fish. Furthermore, one of the most serious problems in tropical man-made lakes is the spread of aquatic weeds, which cut down available oxygen and reduce plankton production. These waterweed invasions not only cause fishery losses but also disrupt lake and river transportation. Damage is often done to irrigation systems and recreation facilities suffer losses through interference with fishing and boating. Weed removal is expensive and sometimes impracticable. On the other hand, limited quantities of these weeds benefit fisheries by providing shelter and feeding grounds and may also have potential as a source of feed for domestic stock.

It is usually considered too expensive to adequately clear forest vegetation from inundation sites, but consequently deoxygenation and toxification of water may result with detrimental effects on the fishery. Nevertheless, despite the many problems encountered in establishing and maintaining a fishery in a man-made lake, there is often great potential for increasing fish production.

Forests inundated by a man-made lake are also an obvious loss of a resource. In addition, the forests in areas downstream from a dam are often adapted to periodic flooding and may be unfavourably influenced after dam completion.

Wildlife values are also affected by river basin development. These basins are among the most productive wildlife areas in Africa because of the availability of much high quality food and the special habitats offered to aquatic or semiaquatic species such as crocodile and hippopotamus. The wildlife populations of the Luangwa valley in Zambia are for example world famous. River basin developments can have several detrimental effects on wildlife. Loss of productive riverine habitat often results. This may imply the virtual elimination of endangered species as in the case of the red lechwe found on the Kafue Flats in Zambia which has been brought about by the construction of the Kafue dam. Loss of important food supplies is common because of alteration of the water regime downstream. Many species of wild ungulates depend for their survival on the availability of flood plain fodder during the dry season. Also, many birds specialize on food organisms produced by intermittent flooding. The open-billed stork, for example, depends on snails which become available after the flats fall dry. Suitable nesting sites for crocodiles tend to disappear because of rapid draw-downs and flooding.

Up till this point reference has been made to the effect of river basin development on existing resources. The other problem that exists is how malpractice of land use in a watershed can affect the life and effectiveness of a reservoir. Inadequate attention is paid to watershed forest conservation and the local population is often allowed to cut, burn and overgraze woodlands in the watershed. This results in accelerated erosion and increased silt loads being carried by streams to the reservoir, reducing its effective life. In addition,

Photo 46. An FAO fisheries expert, with assistant, measures oxygen content of Lake Kariba water, Zambia. Lake Kariba is one of the largest man-made reservoirs, constructed for hydro-electric power and fish production. Photo: UN, issued by FAO.

because the forests do not function effectively as sponges releasing water gradually after periods of rainfall, because the soil has become compacted, much of this water passes over the spillway and is a loss to power production or irrigation requirements.

In general, too little attention is paid in Africa to the protection of the watersheds of man-made lakes against erosion. It is particularly essential that hills or mountains from which water drains into reservoirs are covered by vegetation as much as possible. Cropping on hilly land should therefore be restricted or proper soil conservation practices, such as terracing, strip cropping or planting of cover crops should be maintained. Overgrazing by livestock should obviously not be allowed and proper range management practices should be instigated.

It seems pertinent here to say a few words about the human problems associated with river basin developments, because the human population of a watershed about to be impounded is of course very much concerned. Flood plains are most productive lands and usually the people who are pushed out by inundation are resettled on much less productive sites, and nutritional and related disease problems, social disruption and loss of homes result. The future of these displaced people is in almost all river basin developments inadequately safeguarded. Even in the Lake Volta project, which is generally heralded as one of the greatest success stories of international assistance, the displaced people, although provided with adequate housing, are now occupying land that is not able to feed them adequately.

Prospects in river basin developments

In view of the many problems which have been created by existing river basin developments in Africa, it seems imperative that in the planning of future projects much more attention is given to ecological considerations and that cost/benefit analyses account carefully for the ecological impacts and side effects of various developments. The emphasis should therefore be in the future on multipurpose, integrated planning.

The first step to be taken should be comprehensive evaluations of the ecological and social impacts of planned development. All relevant ecological factors, including fisheries, wildlife, forestry, aquatic weeds and water quality should be carefully assessed. Environmental monitoring should be carried on for all phases of development from the pre-project to the post-project phase. On the basis of this guidelines can be developed under which the impact on the existing natural resources can be minimized and sufficient attention can be given to a well-balanced use and further development of these resources. Studies of the social and cultural systems of the people in a watershed should aim at developing settlement patterns which would be more congenial to an economically developed river basin.

Experience gained with successful multi-purpose river basin developments indicates that these offer a wider range of productive benefits to the community at large than single purpose ones.

Food, health and nutrition

Any discussion of the African environment should also include, as a matter of course, the question of food, health and nutrition and the environmental quality considerations related to these.

With regard to food, I have already indicated that Africa is basically on a collision course. Even under the best of circumstances, keeping in mind the 'Green Revolution' and rapid technological improvements, the food supplies will predictably be inadequate in those parts of Africa where they are most needed. High population increments, low levels of food production and food purchasing power, causing low levels of food consumption, all combine to lessen the chances of improved standards of living and health.

The production of staple foods in Africa is generally low both in quantity and in quality. It is therefore of the greatest importance that traditional cereals such as millets be replaced as much as possible by high-yielding varieties with higher protein content and quality such as cowpeas, groundnuts, soya beans and various high-yielding varieties of rice and maize.

Protein caloric deficiency diseases and other forms of malnutrition, such as kwasikor which is readily recognized in young children by the presence of swollen bellies, spindly legs and reddish-tinged hair, are widespread phenomena in many parts of the continent. The diets of most rural Africans in central, east and west Africa now vary from 5 to 25 grams of animal protein a day with a mean value of 11 grams. With mounting population pressure in rural areas there is an inevitable tendency to change from protein-rich staples to carbohydrate staples.

The hygienic aspects of nutrition have been too often overlooked by policy makers who have increased food production foremost in their minds. Programmes concerned with environmental health have for a long time been limited mainly to activities in water supply and waste disposal, but food quality and hygiene should also be carefully considered. Lack of attention to the latter problems has created health hazards and has resulted in the prevalence of diarrhoeal diseases, food poisoning of infectious or toxic types and chemical food poisoning due to so-called non-intentional additives and pesticides.

Unfortunately, development aid has been instrumental not only in meeting certain major objectives, such as the raising of agricultural productivity, but also in producing other changes which have often fostered the spread of disease or accelerated nutritional deficiencies. These include increased population movements, changes of patterns of water flow and use, and overall changes in man-habitat relationships (e.g. the use of new land for crops and changes in vegetative cover).

New problems have arisen in recent decades as a result of reservoir construction. For example, the flooding of the Kariba reservoir produced an outbreak of sleeping sickness as well as a decline in nutritional standards and overall health status (APTED, *et al.*, 1963).

No doubt the greatest current disease problem caused by man-made lake and irrigation projects is that of schistosomiasis or bilharzia, caused by snail-borne infections. The rapid growth of dense human populations coupled with poor sanitary facilities and increasing mobility of infected people have contributed to the rapid spreading of snail hosts and consequently the serious problem that exists today.

Arthropod-borne infections such as onchocerciasis, trypanosomiasis, filariasis and malaria are also common. The construction of irrigation works has particularly helped to spread the latter. The larval and pupal forms of the small black fly which is the vector of onchocerciasis or river blindness inhabit rapidly flowing streams, and reservoir inundation of these streams has often had the happy effect of reducing the vector's prime habitat but, unfortunately, spilways often provide suitable breeding sites for the fly. Trypanosomiasis has spread as a result of increased mobility of rural people through better means and systems of transportation.

Although foreign aid experts have advocated the construction of fish ponds as one of the easiest and most effective means of providing more cheap protein, it should be kept in mind that they do carry with them health control hazards, such as malaria and schistosomiasis, and should, therefore, be well constructed and managed in order to protect the surrounding population against these potential hazards.

Gross overcrowding and lack of sanitation in many African cities has lowered standards of health and caused the spread of infectious diseases, such as gastroenteritis. Urbanization in general has had a negative effect on the nutritional standards of the people, particularly on migrants from rural areas. The high incidence of nutritional disorders in Africa reflects the seriousness of the social and economic problems facing many Africans in these times.

It can clearly be seen that because many development activities have not been effectively integrated or coordinated with the existing ecosystems, there has been a generally adverse effect on the food and health of the population. Obviously many efforts should be made to rectify this situation. In this regard it must be remembered that we are dealing with a dynamic situation, involving people as well as environmental situations which are constantly changing. Therefore, a balance should be struck between what is desirable and feasible – unfortunately quite a distance lies between the two!

VI. PLANNING FOR THE FUTURE

The need for regional planning

There are people who, unaware of the large cities and population problems which exist, may question the need for regional planning in Africa, which they still consider the continent of wide-open spaces. Yet, Lagos, Ibadan, Accra, Abidjan, Cairo and Johannesburg are all large metropolitan centres whose populations are increasing at a pace which outstrips most other cities of the world.

Unquestionably a need exists for urban planning and along with this an increasing need for regional planning, because the areas around each concentrated settlement are subject to increasingly heavy stresses. DASMANN (1972) has pointed out that cities do not exist apart from their wider environment: food and water flows in, wastes flow out, people move back and forth. The city is merely a focus for activities which encompass a much broader area.

The last few decades have seen the rampant spread of bidonvilles or shantytowns of every type and description around the city edges, whose crowded inhabitants are all too often suffering from ill health, undernourishment and illiteracy. Many people may not realize that Lagos and Ibadan, both with populations exceeding three million people are rapidly turning into the Calcuttas of Africa, with similar problems. These large cities are increasingly filled with throngs of unemployed or under-employed people who have drifted in from the rural areas and who are a threat to the stability of society. The pathology of this situation is a combination of the effects of malnutrition and poverty, and the physical and psychological effects of heavy stresses.

Unhappily, food quality and environmental sanitation have not generally been given the attention they deserve. Health hazards in the form of diarrhoeal diseases and food poisoning have consequently resulted, and epidemics or contagious diseases, as well as the spectre of malnutrition continuously hover as a result of ever-increasing population pressure. Urban priority considerations should include housing, better sanitation, hospitals, schools, nutrition, transportation, etc.

Metropolitan centres naturally require vast amounts of food and water which surrounding areas must provide. This creates employment for the rural population in the vicinity through the production of vegetables, fruits, and other crops. Voluminous outputs of sewage are produced by each city which are traditionally passed into the nearest river, resulting in pollution and more public health hazards affecting the downstream rural population.

In the aggregate considerable portions of the continent are drastically influenced by urban developments and very few unspoiled habitats remain near the larger cities. There are fortunately a few exceptions to this. One outstanding example is Nairobi National Park, situated in a suburb of Nairobi, Kenya and within a 20 minutes driving distance from the international airport. This park is visited annually by hundreds of thousands of people, all anxious to observe the rich savanna fauna of rhino's, giraffes, and herds of zebra, wildebeest, kongoni and other species of antelopes. Similar possibilities, although not so exceptional, exist near other cities.

Many African countries are increasingly making use of development plans. In these plans stress is laid on rural development and employment, self-sufficiency of food, agricultural diversification and the earning power of foreign exchange. Most of these countries have either a long-term strategy or are working on perspective plans to provide guidelines of development over a decade or so from which medium-term plans will evolve. Most have five-year plans, although some are of shorter or longer duration. Usually the main objective is stimulation of rural development.

Unfortunately, programmes of agricultural development are rarely preceded by a necessary evaluation of resources and their use. Comprehensive and integrated land use planning is all too often lacking and where it does exist it does not generally involve the local communities.

On the whole, education, training, research, technical assistance and extension programmes do not adequately meet the various requirements of environmental protection arising from the technological changes in agriculture.

In view of rapidly increasing urban populations and the need to provide essential food supplies, as well as a reasonable quality of life, it is clear that much more emphasis will have to be placed on regional planning in the future.

Ecological considerations in land use planning

Introduction

Land use plans are the best guidelines to wise and improved use of natural resources. They indicate the optimum use of natural resources under the present state of technology and population and include the development of changes from present land use.

Comprehensive land use planning and zoning is a tool which a society can use to help insure its survival and a better life. It employs a system of planning that recognizes the various land resources, a society's needs, and sets forth the coordinated level of use to best meet present and future demands. Land use planning is a relatively new concept in the developing parts of Africa, but it is a badly needed tool. In its absence, many agencies at work in the different sectors of resource conservation are independently developing separate plans, usually limited in scope and often with detrimental results for the country as a whole.

One of the objectives of this type of planning is that lands of high potential for agriculture should be used to meet this potential through the production of crops. As the need for food is high and ever increasing, other forms of land use should, in general, be considered secondary to this main objective. It is for the secondary uses that more detailed assessments of land use are particularly necessary. All those lands of a low potential for agriculture should be assessed in terms of their possible usage for ranching, forest management, wildlife management or multiple land use purposes.

Planning for harmonious development of land use recognizes a unity of nature and man. Such planning is possible only on the basis of a comprehensive appraisal of various environmental issues, particularly economic and ecological.

It is necessary to introduce environmental considerations into our planning and development. Along with effective conservation and rational use of natural resources, protection and improvement of the human environment is vital for national well-being. It is particularly important that long-term basic considerations should prevail over short-term commercial considerations and that social costs and benefits of resource uses be used as the yardstick rather than private gains and losses. Unfortunately, long-term considerations are largely overlooked at present.

There is now an urgent need to pay more attention to the efficiency and long-term stability of modern agricultural and silvicultural systems as ecosystems, particularly in relation to the exhaustion of resources and deterioration of the environment.

Planning for better forms of land use must involve two related objectives: the first is to raise productivity, and the second is to safeguard the higher levels of production attained by adopting appropriate techniques to conserve soils, soil productivity and water. This will be far from easy because of the increasing pressures on the land.

Land use planning designed to maintain ecological balance demands a knowledge and understanding of the environment. This involves study of the effects of different land use practices and cropping systems which may lead to loss of fertility, a reduction in organic matter content, disruption of water balance, increased desiccation and finally to the loss of soil.

In land use planning for natural resources, the regional rather than the national view should be taken and developed to achieve a more balanced development: This is not now the case, particularly in newly independent Africa where planning is done almost entirely at the national level. Efforts made to coordinate planning between adjacent nations under British rule were initially successful in East and Central Africa, but have been weakened or destroyed because of political problems.

For any region a fairly wide range of options in land development and use may exist in relation to the capital input available. For this reason, alternative objectives, using different combinations of options, should be carefully considered and assessed. For example, if a major water system is to be used for hydro-electric power, all the subsidiary effects such as the containment of water in reservoirs, the control of rivers and floods, the development of irrigation systems from such controlled waters, and the effects of such changes on the countryside, agriculture, the life of the human population and on the ecosystems involved should be carefully evaluated and alternate possibilities for development be considered. (see for example discussion about the Aswan dam, p. 209).

Although land use planning is primarily a national responsibility, an important part of the planning process should be carried out at the local level, involving rural communities and taking into account the diversity of the environment.

The need for ecosystem planning

The aim of ecological planning is to arrive at well-balanced, sensible decisions and to avoid the mistakes which can be caused by concentrating on only one aspect of a problem to the exclusion of all others. In terms of land evaluation and utilization the goal is to help in creating and/or preserving useful and stable organizations or ecosystems, in support of the primary human endeavour which should be to live as well as possible on a sustained-yield basis. A self-regenerating rain forest which preserves the fertility of its soil and its genetic composition is a stable ecosystem, but an area of farmland supporting a well-balanced rotation of crops and of livestock, maintaining the fertility of its soil with fertilizers and other techniques and preventing erosion by water and wind is also stable.

There is a justification for maintaining both systems, although the second one is man-made and basically less stable, because of man's continuous interference and his use of energy produced by the ecosystem.

The important consideration to stress here is that each local environment should be regarded as a functioning ecosystem within which agricultural and other land use developments take place and to which these should adapt themselves. To accomplish this, the inter-relationships between ecological conditions, types and intensities of land use and management practices, and problems of environmental degradation should all be carefully studied. Based upon currently available information, steps should be taken to initiate and implement land use planning.

In the future, the findings of basic ecological research should be applied to the study and design of ecologically stable systems of land use and agricultural practices, whereby: a. the productive capacity of land resources can be maintained on a long-term and sustained-yield basis, and b. human, animal and agricultural wastes can be disposed of or recycled in these systems without harmful effects on natural resources or on the environment generally.

Failure to invoke ecological studies and to assess the responses of crops, forest plantations and livestock within the ecological constraints of the various bioclimatic zones are bound to initiate developments either uncertain or disastrous in their consequences.

Control over land use necessitates: 1. surveys to determine land capability and those uses for which an area is best suited, 2. planning to arrange that lands are used rationally in accordance with their capability, and 3. ability of governments or private organizations to direct the movement and settlements of people in accordance with the plan. Land capability can be expressed in terms of carrying capacity, that is the number of individuals of any species of animal that a particular unit of land can support, including human beings. Reasonable progress has been made in this regard in the Republic of South Africa and also in Kenya.

One thing that should be kept in mind in any land use planning in developing countries is that the present form of land use is not necessarily the future form and that in fact as a country develops, patterns of land use are apt to become more varied and intricate. An example of this is, for instance, Rhodesia where until recently the main cash crop was tobacco, but where agricultural production has become greatly diversified, including for instance large-scale production of citrus crops.

Riney (1968) has suggested the use of 'conservation criteria' to determine to what extent present forms of land use are meeting the minimum requirements for conservation. To apply the conservation criterion, appropriate evidence of the present status of the vegetation (including litter, bare ground, grass, shrubs, and tree cover) is needed. Also needed is evidence indicating the present trend in the condition of the vegetation and of potential productivity. When such evidence is related to past and present forms, or intensities of land use, we are obtaining objective evidence facilitating the recommendation of patterns of land use that are at least ecologically sound. Other criteria such as the potential productivity of an area for wildlife or recreation should not be overlooked in land use planning.

Ecological considerations should be an integral part of the development process, as are engineering and economics. The ultimate objective of most development activities should be the improvement of human welfare, not only in terms of food production, but also in terms of environmental quality. If ecological considerations are not adequately taken into account, this objective may not be realized.

An ecologically sound land management plan should begin with a comprehensive ecosystem analysis, including an understanding of energy flow, nutrient cycling, population dynamics and other species relationships crucial to successful land management. For example, it is often necessary to know the reproductive cycle of key plants for successful range management. If certain range plants are heavily grazed before they reproduce, regeneration will be unsuccessful, and the desired species may be replaced and succeeded by ones with lower palatability. It requires considerable effort by well-trained personnel to obtain this type of information and unfortunately so far far too few capable field workers are available to do the necessary work.

Over much of Africa, the association between existing vegetation and land potential remains relatively clear and uncomplicated, because a sufficient number of species representing the original plant associations are still present. Therefore, an ecological method of distinguishing vegetation soil types can be readily used. This has the advantage that the land units distinguished in this way can be related fairly readily to African traditional practices. The basic conception behind such a survey is that the study of vegetation in relation to soil, climate, and other environmental factors should provide the most practical single guide to agricultural and forestry potentialities.

From a knowledge of the natural vegetation it is possible to arrive at a fair estimate of the agricultural potential of almost any area. Natural vegetation is in fact the best indicator, resulting as it does from the sum of the effects of rainfall, soil type, and temperature. Among the different species which are available, there are often a few dominant or characteristic species which can be used as indicators of this potential. Particularly in arid areas there is often a close and clear relationship between the distribution of tree species, rainfall, and soil texture.

Industrial development

Industrial development in Africa frequently has adverse influences on the renewable natural resources and will eventually cause many problems, such as pollution, comparable to those now being faced by the more developed parts of the world. However, so far, with the exception of the extreme north and south of the continent there is virtually no large-scale manufacturing development in Africa.

Most African governments are attaching increasing importance to industrial development and intend to devote an increasing share of available resources for that purpose. One type of industrial development may play a favourable and strategic role in these countries and that is the establishment of relatively small-scale rural industries for the processing of local agricultural products, such as cereals, fruits and vegetables, coffee, cocoa, cotton, palm-oil, hides and skins. Industrial development in most African countries is so far closely linked to agriculture, for farm products are the principal, and in many cases the only raw materials produced by these countries.

The establishment of integrated modern forest industries should also become a major line of development for some countries. They include the manufacturing of plywood, various types of composite building boards and the pulping of wood. In most parts of Africa a serious limiting factor to industrial development is the scarcity of water. As water is already a limiting factor to crop production, and as priority has to be given to the production of food in view of increasing population pressure, obviously water scarcity will hamper the development of those industries which require vast quantities of water in their production process.

In inter-tropical Africa in particular industrial development is still in its early stages. Apart from the processing of export produce, manufacturing activity has as yet gone little beyond the setting up of flour mills, cement plants, soft drink factories, breweries, an occasional cigarette factory and textile mill, and some development in other comparatively simple fields. Many factors inhibit the growth of industrialisation, and economists generally point to the static nature of the rural sector as the most limiting factor of crucial importance.

The market for manufactured goods among a population consisting largely of subsistence cultivators and small commodity producers is far too small to warrant industrial development comparable to that of the more advanced nations, and there can be no significant increase in purchasing power without a very large increase of agricultural output. Yet, any general and substantial increase in agricultural productivity and prosperity depends on the development of the non-agricultural sectors, the creation of jobs for the surplus rural population, and the consequent generation of expanding markets for farm products (ALLAN, 1965).

Without this, the economic situation must remain tied to the few export crops for which bulk markets are available in the indsutrialised countries. The most valuable of these crops (such as cocoa and coffee) are restricted to certain environments by soil and climatic factors. These yield incomes to relatively few producers, and such incomes have only a small diffusion effect.

Although so far little progress has been made in industrial development in tropical Africa, this does not justify a neglect of planning for industries. In fact, if adequate plans are formulated for pollution abatement before industries are established, many of the mistakes that have been made in the developed part of the world could be prevented.

Considering its relatively low population, South Africa is rather industrialized, and some of the problems of the more developed parts of the world are already rearing their ugly heads. Much destruction of rural land has resulted from urban and industrial development and the construction of hydrological projects, roads and railways. Another area where similar problems are looming up is the Nile Delta in Egypt, where considerable hydroelectric power will be available from the Aswan dam for indistrial development.

ENVIRONMENTAL QUALITY CONSIDERATIONS

With the exception of the more advanced parts of northern and southern Africa, so far few Government authorities or leaders of public thinking have given much attention to environmental quality considerations in their respective countries. Most of them would rather consider this as a problem typical of the more advanced parts of the world and not something particularly pertinent to Africa. The prevailing opinion seems to be that if the more developed countries of the world cannot solve their environmental quality problems effectively, why should the people of Africa worry much about it?

Although pollution problems of Africa are still negligible compared with those of Europe or North America, their presence is becoming more apparent. Water pollution, in effect, has already locally assumed serious proportions, such as pollution caused by the mines in Zambia and Zaïre.

Many beaches in West Africa are continuously polluted by oil discharged from boats passing along the coast. The lagoons around Abidjan, Ivory Coast, which has aspirations of becoming 'Africa's riviera', are badly polluted by oil and sewage. Several rivers downstream of metropolitan centres, like Lagos, are also badly polluted mainly with raw sewage, and little effort has so far been made to rectify this situation. Serious water pollution by human wastes, all too often untreated or with scant treatment, frequently results in the spread of water-borne diseases, such as cholera and typhus. In the more advanced parts of Africa water pollution by fertilizers and pesticides is also becoming a distinct problem (see also p. 158).

Air pollution of a chemical nature is largely localized, and mainly restricted to industrial areas in northern and southern Africa. But often forms of air pollution, namely dust storms and the suspension of other particles in the air as a result of fires and erosion are widespread. A night flight taken over East or Central Africa at the start of the dry season is a real education in terms of what man is doing to destroy his environment, for almost everywhere beneath one the land shimmers with the light of unchecked fires.

Because indigenous Africans are so little conscious of environmental quality problems which threaten their continent, there is a real need for education in this field to create greater awareness of the problems that will confront them in the not too distant future. It should be impressed upon them that a healthy environment is vital to a good life and that it is the responsibility of each generation to maintain the productive capacity of land, air, water and wildlife in a manner which leaves successive generations a healthy, productive and enjoyable environment.

The developing countries of Africa at present have the advantage that they are not much afflicted by pollution from industrial waste and their Governments still have the possibility in their long-range forecasting of preparing themselves to use preventative controls against the possible ill-effects of industrialization. While increased production remains essential, the environmental factor has now become a matter of great importance and must be given more attention.

No doubt the recent establishment of the United Nations Environmental Programme in Nairobi will focus more attention on the environmental problem than it has received to date on the continent and it is hoped that this will induce even more leaders of the various countries to pay greater attention to environmental quality considerations.

Ecological constraints to man's future in Africa

An effort has been made to make the reader more aware of the ecological problems which face the people of Africa and to illustrate not only man's detrimental influences on his environment, but also how, with access to modern technology and under improved management, the productivity of agricultural land, range land and forests can be increased.

But what are, or should be, our ultimate aims in increasing the productivity of the land? To feed the people who would otherwise starve to death, so that they can continue to reproduce themselves and raise the population level to an even higher plateau, so that mankind can deplete the remaining resources still further?

Regretfully I must admit that in the many papers I have read about land use in Africa and the need for more technical assistance to step up productivity, I have seen very few that make good sense with regard to the problem of establishing some sort of an ecological balance between man and his environment. One of the more illuminating is a paper by ALLAN (1949) entitled: 'How much land does a man require?', in which he defines the carrying capacity for human beings as follows: 'The carrying capacity of an area is an estimate of the number of people that area will support in perpetuity, under a given system of land usage, without deterioration of land resources'.

The last part of the sentence is the most significant, of course. An area can carry a very much larger population for a number of years than it can support on a continuing basis, but that can only happen through using up the land capital at a greater or lesser rate according to the degree of congestion. There is also a time lag between the beginning of overcrowded conditions and the point at which obvious symptoms of land degradation and soil erosion appear.

The length of this time lag depends on a number of factors, including the type of land and the system used on it, the degree of congestion, the original population density and the rate of population increase.

Once the obvious symptoms of land degradation appear, it is usually too late and too costly to rectify the situation and to make strenuous efforts to bring the land back to good heart, because ecological conditions have become unsatisfactory for the survival of man and beast alike. This is exactly the situation which prevailed in large parts of the Sahelian zone and in Ethiopia in 1973. From this major calamity, let us hope that man will learn his lesson and that concerted efforts will be made to prevent such conditions from reoccuring, perhaps on an even larger scale!

Planning for development: a positive approach to more efficient land use

Faced with tremendously mounting demands for agricultural and forestry products, there is a great need for improved management of the African environment. On some agricultural land there is still a vast untapped potential for increased output. Present levels of production are far below what they could be if the institutional framework were improved, if finances were available and, more particularly, if better techniques were applied.

The adoption of the better techniques in both cash and food crop production can obviously help to raise standards of living, but in most areas methods have not appreciably improved and it would be too much to be expected that in a short space of time the rural population of Africa can be turned into efficient farmers! Planned action is needed to bring an adapted form of technical improvements to African rural life. A new type of agriculture is absolutely imperative, which calls for an integrated approach towards land use planning and development. This implies that various patterns of agriculture be used in combination, rather than in isolation, where such mixed patterns are more profitable.

Land should be classified according to its appropriate potential for arable farming, grazing or forestry, and for each of these major uses suitable types of farming and management practices should be determined. For example, some of the land now used for grazing and cultivation may be better used for tree crops or reforestation, either because such use would be economically more profitable or would prevent erosion and desiccation which may have resulted from past agricultural practices. Often a combination of shelterbelts and tree plantings for local fuel and construction purposes proves very useful. Similarly, programmes for improving livestock production should also be based on a combination of better methods, including better management of existing grasslands and feed crops, and improved breeding and veterinary services.

The determination of the most suitable combination of complementary practices in a given area deserves much greater attention than it has been given so far. Planning for development does not only raise technical issues. There is also the essential educational and political task of explaining to people at all levels the development plans that are proposed and of enlisting their continued enthusiasm and support.

The structure of the various ministries of agriculture generally needs to be improved and integrated. The planned acceleration of agriculture in the framework of economic and social development demands that the various services such as soil conservation, irrigation, animal husbandry and forestry be integrated on the basis of a coherent long-term plan for development.

Because agriculture is so important, and because agricultural backwardness is a major cause of the distressing poverty to be found everywhere, the problem of general education should be reviewed in this context. There are particular difficulties in the way of securing the necessary administrative and executive cadres, whose lack undoubtedly presents one of the greatest problems in Africa today.

Better utilization of food resources, actual and potential, which involves changes in food habits, can be achieved only through more public education. To combat malnutrition on a large scale, there is a need for a very cheap animal protein concentrate, and fish meal and flour in various forms holds out much promise. A move away form the starchy foods to foods with higher protein content is essential. Further product development, combined with large-scale acceptability tests and promotion compaigns, should be undertaken as a matter of urgency. Improved agricultural production techniques require suitable land tenure arrangements. These will have to be devised so as to be in harmony with the stage and context of the economic and social development of the territories concerned. This will demand a great deal of flexibility. In planning for development it is imperative that the various patterns of human impact on ecosystems be studied carefully in order to control the course of economic development with a view to keeping environmental productivity at desirable levels in relation to the quality of the environment.

As western paternalism is more or less finished, the African people, whatever their race or creed, must now overcome current petty nationalisms and decide for themselves what their members will come to be, to what extent they will want to adopt the western world's destructive technology, and what final balance they will want to strike between wilderness and civilization. Hopefully, appropriate decisions will be taken soon, because time is of vital importance!

Acknowledgements

I herewith acknowledge with thanks the criticisms and suggestions of my colleagues H. Goetz and J. M. French.

REFERENCES

Anonymous. FAO Africa survey. Report on the possibilities of African rural development in relation to economic and social growth, FAO, Rome, 168p, 1962.

Anonymous. Report of long-term Agricultural Policy Commission, the mandated territory of South-West Africa. p. 257. roneo. 1949.

Achi, K. Salinization and water problems in the Algerian north east Sahara. *In:* the careless technology, pp. 276–288. The Natural History Press, 1972.

Acocks, J.P.H. Veld types of South Africa. *Bot. Surv. of S.Africa. Mem.* 28, 1963.

Acocks, J.P.H. Karoo vegetation in relation to the development of deserts. *In:* Ecological studies in Southern Africa, Davis, D.H.S. (ed.), pp. 100–125, Monographiae Biologicae, vol. 14, Junk, The Hague, 1964.

Allan, W. How much land does a man require? Rhodes-Livingstone Papers, no. 15, pp. 1–23, 1949.

Allan, W. The African husbandman. Oliver and Boyd, Edinburgh, 505 pp. 1965.

Aubréville, A. The disappearance of the Tropical Forests of Africa. *Unasylva*, 1 :5–11, 1947.

Aubréville, A. Climats, forêts et désertification de l'Afrique tropicale, Soc.d'Edit. Géogr.Marit. et Colon. 351 pp. Paris, 1949.

Apted F.I.C. *et al.* A comparative study of the epidemiology of endemic Rhodesian sleeping sickness in different parts of Africa. *J. Trop. Med. and Hyg.* 66, 1–16.

Bennet, H.H. Soil erosion and land use in South Africa. 1945.

Bourlière, F. & M. Hadley. The ecology of tropical savannas. Annual Review of Ecology and Systematics, 1, pp. 125–152, 1970.

Boulos, L. The discovery of Medemia palm in the Arabian Desert of Egypt. *Bot. Notiser*, 121 :117–120, 1968.

Boyko, H. (ed.) Saline irrigation for agriculture and forestry. Junk, The Hague, 1968.

Brown, L.H. The biology of pastoral man as a factor in conservation. *Biol. Cons.* 3(2) : 93–100, 1971.

Campbell, A.C. & G. Child. The impact of man on the environment of Botswana. Botswana Notes and Records, vol. 3, pp. 91–110, 1971.

Child, G. Wildlife utilization and management in Botswana. *Biol. Cons.* 3(1) : 18–22, 1970.

Child, G. The future of wildlife and rural land use in Botswana. Proc. SARCCUS Symp. Nature and Land Use, Gorongosa Nat. Park, Mozambique. 1971.

Clark, J.D. The prehistory of Southern Africa. Pelican Books, London: Penguin Books, 1959.

Clarke, K. The reconciliation of wildlife conservation with forestry. Proc. Symposium on wildlife management and land use. *E.A. Agricultural & Forestry J.*, vol. 33, pp. 213–217, 1968.

Clatworthy, J.N. A comparison of legumes and fertilizer nitrogen in Rhodesia. Proc. 11th Internat. Grassland Congr., pp. 408–11, 1970.

Clawson, M. Natural resources and industrial development. Johns Hopkins Press, 462 p. 1965.

Cloudsley-Thompson, J.L. The zoology of tropical Africa. The World Naturalist, Weidenfeld & Nicholson, 1969.

Curry-Lindahl, K. Let them live. 399 pp. Morrow, 1972.

Darling, F.F. & M.A. Farvar. Ecological consequences of sedentarization of nomads. *In:* The careless technology, pp. 671–683. Doubleday, 1972.

Darling, F.F. Wildlife husbandry in Africa. *Sci. Amer.* 203:123–128, 1960.

Darlington, P.J. jr. Zoogeography: The geographical distribution of animals, 675 pp., Wiley & Sons, 1963.

Dasmann, R.F. African game ranching. MacMillan, New York, 75 pp. 1964.

Dasmann, R.F. Environmental Conservation. Wiley and Sons Ltd. London. 473 p. 1972.

Dasmann, R.F. *et al.* Ecological principles for economic development. Wiley and Sons Ltd. London, 1973.

DASMANN, R.F. A system for defining and classifying natural regions for purposes of conservation. IUCN Occasional Paper no. 7, mimeo, 47pp., 1963.
DAVIDSON, R.L. An experimental study of succession in the Transvaal highveld. *In:* Ecological studies in southern Africa. DAVIS, D.H.S. (ed.), pp. 113–125, Monographiae Biologicae vol. 14, Junk, The Hague 1964.
DAUBENMIRE, R. Ecology of fire in grasslands. *In:* Advances in Ecological Research. V, pp. 209–266. Acad. Press. London, 1968.
DAVIS, D.H.S. Notes on the status of the American grey squirrel (*Sciurus carolinensis* Gmelin) in the south western Cape (South Africa) *Proc. Zool. Soc. London,* 120:265–268. 1950.
DAVIS, D.H.S. *et al.* (eds.) Ecological studies in Southern Africa. Monographiae Biologicae 14; 415 p. Junk, The Hague, 1964.
DORST, J. Avant que nature meurt, Neuchâtel. Delachaux et Niestlé. 424 p., 1965.
DREGNE, H.E. ed. Arid lands in transition. AAAS, Washington, DC. 524pp. 1970.
DUMONT, R. False start in Africa. Sphere Books Ltd., London, 271 pp., 1968.
DYNE, VAN G.M. The ecosystem concept in natural resource management. 283p., Academic Press, N.Y. 1969.
ELLERMAN, J.R. & T.C.S. MORRISON-SCOTT. Checklist of Palaearctic and Indian Mammals: 1758 to 1946. British Museum. London. 1951.
ELTON, C.S. The ecology of invasions by animals and plants. Methuen, London, 181 pp. 1958.
EVERAARTS, J.M. *et al.* Contribution a l'étude des effets sur quelques élements de la faune sauvage des insecticides organo-chlorés utilisés au Tchad en culture cotonnière. *Cot. Fib. Trop.,* vol. 26(4) 385–394, 1971.
FAO Report of the conference on the establishment of an agricultural research programme on an ecological basis in Africa. Sudanian Zone. 127 pp. Rome. 1969.
FAO World Indicative Plan for the development of agriculture until 1975 and 1985. Provisional region study no. 3, Africa. Mimeo. 1968.
FAO The state of food and agriculture 1972. FAO, Rome, 189 pp., 1972.
FARVAR, M.T. & J.P. MILTON. The careless technology. Ecology and international development. 1030 p. Doubleday. 1972.
FISHER, J. *et al.* The red book, Wildlife in danger. Collins, 368 pp. 1969.
FISHWICK, R.W. Sahel and Sudan zone of northern Nigeria, north Cameroons and the Sudan, pp. 59–83. *In:* Afforestation in arid zones. Kaul, R.N. (ed), Monographiae Biologicae vol. 20, Junk, The Hague 1970.
FOURNIER, F. & J. D'HOORE. Carte du danger d'érosion en Afrique au sud du Sahara 1:10.000.000. Commission de Co-opération Technique en Afrique. Paris, 1962.
FOURNIER, F. The soils of Africa. *In:* A review of the natural resources of the African continent. Ed. by Unesco, pp. 221–248, 1963.
FOURNIER, F. Aspects of soil conservation in the different climatic and pedologic regions of Europe. Nature and environment series, Council of Europe, 194pp. 1972.
FRYER, G. Some hazards facing African lakes. *Biol. Cons.* 4(4): 301–302, 1972.
FURON, R. The gentle little goat: arch despoiler of the earth. Unesco Courier. vol. 11, no. 1, 1958.
GHABBOUR, S.I. Some aspects of conservation in the Sudan. *Biol. Cons.* 4(3): 228–230, 1972.
GIFFARD, P.L. The use of *Acacia albida* in the rejuvenation of the soil in the dry tropical zones. Proc. 6th World Forestry Congr., III: 3195–3197, 1966.
GLOVER, P.F. A review of recent knowledge on the relationships between tsetse fly and its vertebrate hosts. IUCN Publ. New series No. 6, 84pp. 1965.
GOUROU, P. The tropical World. J. Wiley, New York, 1966.
HAILEY, L. An African survey, revised 1956. Oxford University Press, 1674 pp., 1957.
HASKELL, P.T. Locust control: ecological problems and international pests. *In:* The careless technology, pp. 499–527. Doubleday, 1972.
HARROY, J.P. Afrique: Terre qui meurt. Marcel Hayez. Brussels, 1949.
HAUMAN, L. La région afri-alpine en phytogéographie centro-africaine. *Webbia,* XI pp. 467–9, 1955.
HEDBERG, I. & O. HEDBERG (Eds.) Conservation of vegetation south of the Sahara. Acta Phytogeogr. Suec. 54, 320pp. 1968.

HESSE, P.R. Some facts and fallacies about termite mounds. Tanganyika Notes and Records, no. 39, pp. 16–25, 1955.

HOPKINS, B. *J. Ecol.* 54:687–703, 1966.

HUMPHREY, N. The Kikuyu Lands, Nairobi: Government Printer, 1945.

IUCN Publ. New series No. 4. The ecology of man in the tropical environment. Proc. and papers. Ninth technical meeting. Nairobi. Sept. 1963, 335 pp. 1964.

KASSAS, M. Desertification versus potential for recovery in circumsaharan territories. Proc. international conf. on arid lands in a changing world. Tucson (Ariz). In: 'Arid lands in transition'. Publ. no. 90 AAAS, Wash: 123–142, 1970.

KAUL, R.N. (ed.) Afforestation in arid zones. Monographiae Biologicae 20, 435 p. Junk, The Hague, 1970.

KEAY, R.W.Y. Vegetation map of Africa south of the tropic of Cancer. Explanatory notes by KEAY, 23 pp, Oxford University. Press, 1959.

KOECHLIN, J. Region south of the Sahara. *In:* Natural resources of the African Continent pp. 271–281. Unesco. 448 pp. 1963.

LANGDALE-BROWN, L. *et al.* The vegetation of Uganda and its bearing on land use. 159 pp., Govt. Printer, Entebbe, 1964.

LAMBRECHT, F.L. The tsetse fly: a blessing or a curse? *In:* The careless technology. Ecology and international development. Nat. History press. Garden City N.Y. pp. 726–741. 1972.

LEBRUN, J. La végétation de la plaine alluviale au Sud du Lac Edouard. Inst. Parcs Nat. Congo Belge, Explor. Parc Nat. Alb., Miss. Lebrun (1937–1938) I, 800 pp. 1947.

LE HOUÉROU, H.N. North Africa, past, present and future, *In:* Arid lands in transition. Publ. no. 90, AAAS, Wash.: pp. 227–278, 1970.

LOOSLI, J.K. & V.A. OYENUGA. Role of livestock in developing countries with emphasis on Nigeria. *Fed. Proc.*, 22: 141–144, 1963.

LOWE, C.H. Fauna of desert environments with desert disease information. *In:* Deserts of the world. Univ. of Arizona Press. pp. 569–645, 1970.

LYALL-WATSON, M. Men and animals: their fruitfull co-existence in Africa. *Optima*, vol. 15. pp. 85–94, 1965.

MARTIN, P.S. Prehistoric overkill. *In:* Man's impact on environment. DETWILES, (ed.) pp. 612–625, McGraw-Hill, 1971.

MCGINNIES, W.G. *et al.* (eds.) Deserts of the world. An appraisal of research into their physical and biological environments. Univ. of Arizona Press. 788 pp. 1970.

MOLSKI, B.A. Shelterbelt regions in Nigeria. Proc. 6th World Forestry Congr., vol. III, pp. 3797–3803, 1966.

MOREAU, R.E. The bird faunas of Africa and its islands. Academic Press. N.Y. 424 pp. 1966.

MOREL, G. Contribution à la synécologie des oiseaux du Sahel sénégalais. Mém. ORSTOM, 29, 179 pp. 1968.

NASH, T.A.M. Africa's Bane: Tsetse Fly. London, Collins. 224 pp. 1969.

OEDEKOVEN, K.H. The United Arab Republic. *In:* Afforestation in arid zones. Kaul, R.N. (ed), Monographiae Biologicae vol. 20, pp. 86–97. Junk, the Hague, 1970.

OGBE, G.A.E. Regeneration practices in the high forests of Nigeria. Proc. Sixth World Forestry Congr., Madrid 1966, III, pp. 3167–3169. 1966.

OWEN, D.F. Animal ecology in Tropical Africa. Oliver & Boyd, Edinburgh, 122 pp. 1966.

PETRIDES, G.A. & SWANK, W.G. Population densities and the range carrying capacity for large mammals in Queen Elizabeth Nat. Park, Uganda. *Zool. Afr.* 1:209–25, 1965.

PHILLIPS, J.F.V. Agriculture and ecology in Africa. Faber and Faber, 424 pp. London. 1959.

PHILLIPS, J.F.V. Fire-as master and servant: its influence in the bioclimatic regions of Trans-Saharan Africa. Proc. Tall Timbers Fire Ecology Conf.: 7–109, 1965.

POLUNIN, N. Introduction to plant geography and some related sciences. McGraw Hill, 1960.

POSSELT, J. The domestication of the eland. *Rhodesian J. of Agric. Res.* 1, p. 81–88, 1963.

PRESCOTT, J.R.V. Overpopulation and overstocking in the native areas of matabeleland. *In:* People and land in Africa south of the Sahara. Readings in Social Geography. R.M. Prothero (ed.), pp. 239–253 Oxford Univ. Press, New York, 1972.

Prothero, R.M. People and land in Africa south of Sahara. Readings in Social Geography. Oxford U. Press, 1972.
Rakhmanov, V.V. Role of forests in water conservation. Translated from Russian. Israel Program for Scientific Translations. Jerusalem. 192pp, 1966.
Riney, T. Criteria for land use planning *E.A. Agric. & Forestry J.*, vol. 33, pp. 34–38, 1968.
Russell, E.W. The impact of technological developments on soils in East Africa. *In:* The careless technology. Doubleday, 1972.
Sanford, W.W. Conservation des Orchidées en Afrique occidentale. *Biol. Cons.*, 3(1): 47–52, 1970.
Semple, A. Grassland improvement in Africa. Biol. Cons. 3(3): 173–180, 1971.
Schnell, R. Végétation et flore de la region montagneuse du Nimba. Dakar. (Mem. IFAN no. 22), 1952.
Scott, J.D. Conservation of vegetation in south Africa. Commonwealth Bur. Past. & Field Crops, Bul. XVI, pp. 9–27, 1951.
South Africa, Union of Report of the Desert Encroachment Committee. Pretoria. Government Printer. p. 27 (undated).
Stebbing, E.P. The threat of the Sahara. *Journ. Roy. African Society*, extra suppl. to vol. 36, 1937.
Stebbing, E.P. The creeping desert in the Sudan and elsewhere in Africa. McCorquodale, (ed.) Sudan Ltd: Khartoum 165 pp. 1953.
Stewart, Ph. *Cupressus dupreziana*, threatened conifer of the Sahara. *Biol. Cons.* 2(1): 10–12, 1969.
Stobbs, T.A. Beef production from pasture leys in Uganda. *J. Brit. Grassl. Soc.* 24. pp. 81–87, 1969.
Talbot, L.M. Wild animals as a source of food. Spec. Sci. Rep. Wildlife no. 98. US Dept. of the Interior, US Govt. Printing Off. 16p. 1966.
Taylor, C.R. The eland and the oryx. *Sc. American*, vol. 220(1): 89–95 pp. 1969.
Trapnell, C.G. & J.N. Clothier. Soils, vegetation and agricultural systems of North Western Rhodesia. Lusaka Govt. Printer, 1937.
Thomas, M.E. & G.W. Whittington. Environment and land use in Africa. London, Methuen, 544 pp. 1969.
Rattray, J.M. The grass cover of Africa. *FAO Agric. Stud.* 49:1–168, Rome, 1968.
Richter, W. von, Wildlife and rural economy in S.W. Botswana. Botswana notes and records 2, pp. 85–94, 1970.
Richter, W. von, C.D. Lynch & T. Wessels. Status and distribution of the larger mammal species on farmland in the Orange Free State. Orange Free State Prov. Admin. Res. Report no. 1, 14pp., mimeo, 1972.
Unesco A review of the natural resources of the African continent, 437 p. Unesco, Paris, 1963.
UN Ground water in Africa. Document ST/ECA/147, 1973.
Uvarov, B.P. Development of arid lands and its ecological effects on their insect fauna. *In:* The problems of the arid zone, Proc. of the Paris symposium. Unesco. Paris. *Arid Zone Research* 8: 164–198, 1962.
Verdcourt, B. Why conserve vegetation? *In:* Conservation of vegetation south of Sahara, Acta Phytogeogr. Suec. 54, 320pp., 1968.
Vos, A. de The need for nature reserves in East Africa. *Biol. Cons.* 1(2): 130–134, 1969.
Vos, A. de Wildlife production in Africa. *In:* Proc. III World Conference on Animal Production, Melbourne 1973. Sydney University Press, Sydney, 1974.
West, O. The ecological impact of the introduction of domestic cattle into wildlife and tsetse areas of Rhodesia. *In:* The careless technology (Farvar & Milton eds.), pp. 712–725, 1972.
Whyte, R.O. The use of arid and semi-arid land. *In:* Arid lands. A geographical appraisal. Methuen Paris, 461 pp. 1966.

AUTHOR INDEX

A
ACHI, K., 207
ACOCKS, J. P. H., 97–100
ALLAN, W., 104, 105, 112, 125, 129, 130, 161, 221, 223
APTED, F. I. C., 213
AUBRÉVILLE, A., 134, 177

B
BENNETT, H. H., 100
BOULOS, L., 59
BOURLIÈRE, F., 29, 32
BOYKO, H., 204
BROWN, L. H., 128, 129, 170

C
CAMPBELL, A. C., 94, 95
CHILD, G., 94–96, 192
CLARK, J. D., 102
CLARKE, K., 144
CLATWORTHY, J. N., 172
CLAWSON, M., 119
CLOUDSLEY-THOMPSON, J. L.,
CLOTHIER, J. N., 112
CURRY-LINDAHL, K.,

D
DARLING, F. F., 128, 130
DARLINGTON, P. J., 16
DASMANN, R. F., 28, 190, 215
DAUBENMIRE, R., 111
DAVIDSON, R. L., 90
DAVIS, D. H. S., 137
DEVRED,
D'HOORE, J., 124
DORST, J.,
DREGNE, H. E.,
DUMONT, R., 169
DYNE, VAN G. M., 29

E
ELLERMAN, J. R., 137
ELTON, G. S., 135
EVERAARTS, J. M., 159
FARVAR, M. T., 128, 130
FOURNIER, F., 20, 124, 126
FRYER, G., 187
FURON, R., 155

G
GHABBOUR, S. I., 134
GIFFARD, P. L., 180
GLOVER, P. F., 149
GOUROU, P., 146

H
HADLEY, M., 29, 32
HAILEY, L., 98
HARROY, J. P., 124
HASKELL, P. T., 152
HAUMAN, L., 16
HESSE, P. R., 146
HOPKINS, B., 31
HOUÉROU, LE H. N., 61, 63, 66, 127, 134
HUMPHREY, N., 114

K
KEAY, R. W. Y., 22, 26
KOECHLIN, J., 16

L
LANGDALE-BROWN, L., 123
LAMBRECHT, F. G., 151
LEAKEY, 102
LEBRUN, J., 26
LOOSLI, J. K., 174
LOWE, C. H., 152
LYALL-WATSON, M., 173

M
MARTIN, P. S., 102
MCGINNIES, W. G.,
MOLSKI, B. A., 142
MOREAU, R. E., 67, 103
MOREL, G., 31
MORRISON-SCOTT, T. C. S., 137

O
OGBE, G. A. E., 177
OUDEKOVEN, K. H., 62
OYENUGA, V. A., 174

P
Petrides, G. A., 75
Phillips, J. F. V., 97, 111, 122, 167
Posselt, J., 193
Polunin, N., 136
Prescott, J. R. V., 79
Prothero, R. M., 85

R
Rakhmanov, V. V., 142
Richter, von W., 92, 94, 96
Riney, T., 219
Russell, E. W., 122

S
Sanford, W. W., 144
Schnell, R., 16
Schürholz, 137
Scott, J. D., 91
Semple, A., 105, 173, 175
Stebbing, E. P., 134
Stewart, Ph., 59
Stobbs, T. A., 173
Swank, W., 75

T
Talbot, L. M., 192
Taylor, C. R., 187
Trapnell, C. G., 112

U
Uvarov, B. P., 152

V
Verdcourt, B., 74
Vos, de A., 144, 192

W
West, O., 79, 81, 89, 104
Whyte, R. O., 130

SUBJECT INDEX

A
Abies, 61
Acacia, 49, 68, 69, 71, 77, 85, 86, 92, 93, 100, 134
A. albida, 30, 50, 179, 180
A. ataxacantha, 49
A. laeta, 49
A. mellifera, 49
A. nilotica, 179
A. raddiana, 49
A. senegal, 49, 63, 179
A. seyal, 49
A. tortilis, 179
addax, 17
A. nasomaculatus, 17
Aepyceros melampus 193
African rice, 34
Afromosia etala, 142
Aleppo pine, 61, 63
alfa grass, 63
Ammotragus lervia, 17
Anacardium occidentalis, 179
Andropogon, 41
Anogeissus leiocarpus, 179
Antidorcas marsupialis, 92
Arachis hypogaea, 41
Araucaria, 178
Atriplex, 50, 66
Azadirachta Indica, 42, 180

B
Baikiaea plurijuga, 83
Bambara nut, 117
banded duiker, 39
barbary sheep, 17
Bauhinia rufescens, 49
beans, 41
blackbass, 186
black duiker, 39
black fly, 159
blesbok, 92
bluebuck, 139
bongo, 18
bontebok, 100
Boocercus euryceros, 18
Borassus, 77
Boscia albitrunca, 93
Brachystegia, 22, 69, 71, 77
buffalo, 17
Burchell zebra, 139

C
cacti, 66
Cajanus cajan, 117
Callitris, 61
Cambarra groundnuts, 41
Carica papaya, 38
carob, 180
carp, 186
cashew tree, 179
cassava, 38
cattley guava, 38
Cedrela odorata, 178
Celtis, 69
Cephalophus jentinki, 39
C. niger, 39
C. zebra, 39
Ceratonia siliqua, 180
Cercopithecus mona, 18
Cerobania siliqua, 62
chevrotain, 39
chimpanzee, 18
cist, 63
Cistus libanitis, 63
Citrullus vulgaris, 93
Colobus verus, 39
Colocasia antiquorum, 38
Colophospernum mopane, 24, 77
Combretum, 49, 77
Commiphora, 49, 69, 75, 85, 93
common carp, 137
Congo jute, 38
Connochaetes gnou, 89, 92
C. taurinus, 75
cowpea, 41, 117
Crocidura, 17
Cupressus, 76
C. dupreziana, 59
Currajong, 180
Cypressus lusitanica, 178
Cyprinus carpio, 137

D
Dacryodes guava, 38
Dama dama, 137
Damaliscus dorcas, 92, 100
D. korrigum, 75
Desert locust, 151
Desmodium contortum, 40
dika nut, 36
Dioscorea alata, 38

E
Echinochloa, 77
Egyptian clover, 117
Eichornia crassipes, 136
Elaeis guineensis, 34
eland, 75
elephant, 17
elephant grass, 72
Eleusine, 117
Enantophragma, 69
Equus quagga burchelli, 139
E. quagga quagga, 139
E. zebra, 100
Eragrostis abyssinica, 117
Eucalyptus, 68, 71, 83, 92, 178

F
Felis libyca 17
Felis serval, 192
fennec, 17
Fennecus zerda, 17

G
Gazella rufina, 139
G. thomsoni, 75
giant forest hog, 39
giant sable, 83
Gleditsia triacanthos, 180
Glossina sp., 148
Gorilla g. gorilla, 39
gray aquirrel, 137
grey rhebuck, 92
Groundnuts, 41
guava, 180
Guibourtia coleoperma, 83

H
Hakea, 100, 136
hare, 17
Helichrysum, 25
henbane, 59
Heterotis, 186
Hevea brasiliensis, 38
Hippotragus equinus, 54
H. leucophaeus, 139
H. variani, 83
honey locust, 180
Hyaena hyanea, 17
Hyemoschus aquaticus, 39
Hylochoerus meinartzhageni, 39
Hyoscyanus muticus, 59
Hyparrhenia, 41
Hypericum 69
Hyphaene, 77
Hystrix, 17

I
impala, 193
Irvingia gabonensis, 36
Isoberlinia, 22, 41

J
Jentink's duiker, 39
Julbernardia, 22, 77
Juniperus, 61
J. macrocarpa, 71
J. phoenicea, 63
J. procera, 69, 71

K
Kafue lechwe, 83
Khaya, 69
Kobus leche, 83

L
Lannea humilis, 49
Lantana salviaefolia, 136
leopard, 17
Libyan cat, 17
Lobelia, 25, 69

M
mahogany, 69
Mangifera indica, 38
Medemia argun, 59
medemia palm, 59
Mesquite, 179
Micropterus dolomieni, 137
M. salmoides, 137
Millet, 41
Miraculous berry, 36
Mlanjé cypress, 83
Mona monkey, 18
Manihot utilissima, 38
Mopane, 24, 77
Mountain reedbuck, 92
Musa paradisiaca, 120
M. sapientum, 120
Myocastor coypus, 137
Neem tree, 42

Neotragus pygmaeus, 39
Nomadacris septemfasciata, 151
nutria, 137

O
oil bean, 36
oil palm, 34
okapi, 39
O. johnstoni, 39
Olea chrysophylla, 71
olive colobus, 39
Opuntia 136
O. ficus-indica, 66
oryx, 193
O. beisa, 193
Oryza glaberrima, 34

P
palm tree, 34
Panicum gracilis, 40
Pan troglodytes, 18
papaya, 38
Para rubber, 38
Parkia clappertonia, 30
Paraechinus deserti, 17
Pelea capreolus, 92
Pennisetum, 41, 72, 73, 117
Pentaclethea macrophylla, 36
Phacochoerus aethiopicus, 54
Phaseolus spp., 41
Phoenicean juniper, 63
pigeon pea, 117
Pinus, 83
P. halepensis, 61, 63
P. merkusii, 178
P. radiata, 178
Podocarpus, 69, 85
porcupine, 17
Potamochoerus porcus, 39
Prosopis chilensis, 179
Protea, 90
P. caffra 90
Psidium guajava, 38
Pterocarpus, 49
P. angolensis, 83

Q
Quelea, 46, 57
quagga, 139

R
Raphia sp., 34
red Lechwe, 157, 210
red locust, 151
red mongoose, 17
Redunca fulvorufula, 92
rhinoceros beetle, 36
Rhodesian mahogany, 83
Rhodesian teak, 83
roan, 54
Rosmarinus spp., 63
rosemary, 63
Royal antelope, 39
rufous gazelle, 139
Rynchophorus phoenicas, 36

S
Saharan cypresses, 59
Salmo gairdnezi, 137
S. trutta, 137
Saltbushes, 66
Salvelinus fontinalis, 137
Salvinia auriculata, 136
Schistocerca gregaria, 151
Schweberia mexicana, 180
Sciurus carolinensis, 137
Senecio, 25, 69
serval, 192
Shea Butter, 30
shrew-mice, 17
Simulium, 159
sorghum, 41
Sorghum vulgare, 41, 117
springbok, 92
Stiloxanthus, 40
Stipa tenacisima, 63
striped hyaena, 17
Suncus, 17
Sus scrofa, 137
Sterculia diversifolia, 180
Synsepalum dulcificum, 36

T
taro, 38
Taurotragus oryx, 75
Teak, 178
Tectona grandis, 178
Terminalia, 77
T. ivorensis, 178
T. superba, 178
Tetreclinis articulata, 63
Themeda, 73
Thomson's gazelle, 75
Tilapia, 186, 187
topi, 75
Trifolium alexandrinum, 117
Trypanosomiasis, 187
tsetse fly, 148

U
Urena, 38

V
Varsis, 76
Vigna sinensis, 41
V. unguiculata, 117
Voandzeia subterranea, 41, 117
Vossia, 77

W
warthogs, 54
water fern, 136
water hyacinth, 136
Welwitschia Bainesi, 16
western bush pig, 39
white-tailed gnu, 89, 92
Widringtonia whytei, 83
wildebeest, 75
wild boar, 137

Y
yam, 38